ENERGY SCIENCE, ENGINEERING AND TECHNOLOGY

OIL, GAS, AND COAL RESOURCES ON FEDERAL LANDS

BACKGROUND AND FAIR VALUE ISSUES

Energy Science, Engineering and Technology

Additional books in this series can be found on Nova's website under the Series tab.

Additional e-books in this series can be found on Nova's website under the e-book tab.

ENERGY SCIENCE, ENGINEERING AND TECHNOLOGY

OIL, GAS, AND COAL RESOURCES ON FEDERAL LANDS

BACKGROUND AND FAIR VALUE ISSUES

MARYBETH T. BALDWIN
EDITOR

nova publishers
New York

For permission to use material from this book please contact us:
Telephone 631-231-7269; Fax 631-231-8175
Web Site: http://www.novapublishers.com

Library of Congress Cataloging-in-Publication Data

ISBN: 978-1-63117-645-6

Published by Nova Science Publishers, Inc. † New York

CONTENTS

PREFACE

Production of oil and natural gas from leases on federal lands and waters is an important part of the nation's energy portfolio and a significant source of revenue for the federal government. This book examines steps Interior has taken to ensure the public receives a fair return on oil and gas resources since 2007. It also examines the number of tracts leased, along with the trends in associated coal production and revenues generated since 1990; BLM's implementation of the process to estimate fair market value for coal leases; the extent to which BLM considers coal exports and domestic coal reserve estimates when estimating fair market value; and the extent to which BLM communicates information on federal coal lease sales to the public.

Chapter 1 - In fiscal year 2012, companies received over $66 billion from the sale of oil and gas produced from federal lands and waters, and they paid $10 billion to the federal government for developing these resources according to the Department of the Interior. The federal government seeks a fair return on its share of revenue from leasing and production activities on federal lands and waters through the federal oil and gas fiscal system. Under the fiscal system, companies pay royalties, rents, and other payments—payments generally specified in lease terms—and taxes on profits from the sale of oil and gas produced from federal leases. In May 2007, GAO found, based on several studies, that the government received one of the lowest percentages of value of oil and gas produced in the world. In September 2008, GAO found that Interior had not evaluated the federal oil and gas fiscal system for over 25 years and recommended that a periodic assessment was needed.

GAO was asked to review Interior's collection of oil and gas revenues. This report examines steps Interior has taken to ensure the public receives a fair return on oil and gas resources since 2007. GAO reviewed applicable law and regulations; examined prior GAO studies, Interior policies and documents; and interviewed officials.

Chapter 2 - In fiscal year 2012, about 42 percent of the 1.05 billion tons of coal produced in the United States came from coal tracts leased under the federal coal leasing program. Interior's BLM is responsible for managing this program, including estimating the fair market value of the coal to be leased. GAO was asked to examine this program. (Representative Markey originally made this request as Ranking Member of the House Committee on Natural Resources. He is now a member of the United States Senate.) This report examines (1) the number of tracts leased, along with the trends in associated coal production and revenues generated since 1990; (2) BLM's implementation of the process to estimate fair market value for coal leases; (3) the extent to which BLM considers coal exports and domestic coal reserve estimates when estimating fair market value; and (4) the extent to which BLM communicates information on federal coal lease sales to the public. GAO analyzed data on coal leasing activity, examined regulations and case files for coal lease sales, and interviewed BLM and other officials.

In: Oil, Gas, and Coal Resources …
Editor: Marybeth T. Baldwin

ISBN: 978-1-63117-645-6

Chapter 1

OIL AND GAS RESOURCES: ACTIONS NEEDED FOR INTERIOR TO BETTER ENSURE A FAIR RETURN*

United States Government Accountability Office

WHY GAO DID THIS STUDY

In fiscal year 2012, companies received over $66 billion from the sale of oil and gas produced from federal lands and waters, and they paid $10 billion to the federal government for developing these resources according to the Department of the Interior. The federal government seeks a fair return on its share of revenue from leasing and production activities on federal lands and waters through the federal oil and gas fiscal system. Under the fiscal system, companies pay royalties, rents, and other payments—payments generally specified in lease terms—and taxes on profits from the sale of oil and gas produced from federal leases. In May 2007, GAO found, based on several studies, that the government received one of the lowest percentages of value of oil and gas produced in the world. In September 2008, GAO found that Interior had not evaluated the federal oil and gas fiscal system for over 25 years and recommended that a periodic assessment was needed.

* This is an edited, reformatted and augmented version of the United States Government Accountability Office publication, GAO-14-50, dated December 2013.

GAO was asked to review Interior's collection of oil and gas revenues. This report examines steps Interior has taken to ensure the public receives a fair return on oil and gas resources since 2007. GAO reviewed applicable law and regulations; examined prior GAO studies, Interior policies and documents; and interviewed officials.

What GAO Recommends

GAO recommendations include that Interior establish documented procedures for (1) periodically assessing the fiscal system and (2) determining whether and how to change new offshore lease terms. Interior concurred with GAO's recommendations.

What GAO Found

Interior has taken some steps intended to help ensure a fair return on federal oil and gas resources but does not have documented procedures for periodically conducting assessments of the fiscal system. Specifically, Interior has taken the following steps:

- *Changed offshore lease terms and considered but has not changed onshore lease terms.* Interior changed certain offshore lease terms—including raising royalty rates twice in response to changing market conditions. For onshore resources, which are subject to many of the same market conditions, Interior has considered but not made changes to royalty rates. Interior officials are currently unable to make timely adjustments to onshore royalty rates. Current regulations generally provide for a fixed onshore royalty rate that limits Interior's flexibility to make timely adjustments.
- *Contracted for studies of various aspects of the fiscal system.* Interior contracted for three studies examining its fiscal system including a study done in 2011, in response to GAO's September 2008 report that compared the U.S. government's oil and gas fiscal system to other resource owners. Interior officials said the reports provided some useful information such as how fiscal terms in the United States compared to other resource owners.

- *Interior is examining potential regulatory changes that could simplify royalty payments.* Interior is examining potential regulatory changes that could simplify royalty payments. GAO found in the past that complex valuation regulations can result in inaccurate royalty payments made by industry, and this could increase costs to ensure accurate royalty payments because of the need for potentially detailed and time- consuming audits of records. In May 2011, Interior published the Advance Notice Of Proposed Rulemaking for a proposed rule currently undergoing internal review. According to officials, the proposed rule is expected to be published in 2014, and officials explained that it took several years due to factors including the complexity of oil and gas valuation.

Interior does not have documented procedures in place for determining when to conduct periodic assessments of the fiscal system. Although Interior recently contracted for such an assessment, it was the first in well over 25 years. Without documented procedures, Interior will not have reasonable assurance that it will consistently conduct such assessments in the future and, without periodically conducting such assessments, Interior cannot know whether there is a proper balance between the attractiveness of federal leases for investment and appropriate returns for federal oil and gas resources, limiting Interior's ability to ensure a fair return. Further, Interior does not have documented procedures for determining whether and how to make changes to new offshore lease terms. Without documented procedures for determining whether and how to make changes to new offshore lease terms, Interior's rationale is not transparent and may result in inconsistent decisions. Such inconsistencies would undermine Interior's credibility and ability to better ensure a fair return on federal oil and gas resources.

ABBREVIATIONS

BLM	Bureau of Land Management
BOEM	Bureau of Ocean Energy Management
MMS	Minerals Management Service
ONRR	Office of Natural Resources Revenue

December 6, 2013

The Honorable Ron Wyden
Chairman
Committee on Energy and Natural Resources
United States Senate

Dear Mr. Chairman:

Production of oil and natural gas from leases on federal lands and waters is an important part of the nation's energy portfolio and a significant source of revenue for the federal government. Domestic and foreign companies received over $66 billion from the sale of oil and gas produced from federal lands and waters in fiscal year 2012, according to the Department of the Interior. Interior reported collecting about $9.7 billion in 2012 from royalties and other payments from these companies, making oil and gas resources one of the federal government's largest nontax sources of revenue. The terms and conditions under which the government collects these revenues are referred to as the "oil and gas fiscal system" and generally include royalties and other payments for the rights to explore, develop, and sell oil and gas resources. However, over the past several decades, we, and others, have identified problems with Interior's management of the federal oil and gas fiscal system. For example, in 1982, a task force convened by Interior found that management of the fiscal system needed a thorough overhaul and provided 60 recommendations for improving the fiscal accountability of the nation's onshore and offshore resources.[1] Upon the completion of the task force's work, the Secretary of the Interior informed Congress, in March 1983, that Interior had refined the system and that a "full and fair return" to the American people would be assured. In May 2007, we found that, based on the results of a number of studies, the government receives one of the lowest government takes—commonly understood to be the total revenue, as a percentage of the value of oil and natural gas produced—in the world.[2] In addition, in September 2008, we found that Interior did not routinely evaluate the federal oil and gas fiscal system and suggested that Congress should consider directing the Secretary of the Interior to (1) convene an independent panel to conduct a comprehensive review of the oil and gas fiscal system and (2) establish procedures to periodically evaluate the state of the fiscal system.[3] In 2011, in part because of the challenges identified in our past work concerning Interior not having reasonable assurance that it is collecting its share of revenue from oil and gas produced on federal lands, we added Interior's management of federal oil and gas resources to GAO's list of programs at high risk of fraud, waste, abuse, and mismanagement.[4]

Interior has oversight responsibility for the development of federal oil and gas resources located under over 260 million surface onshore acres, 700 million subsurface onshore acres, and more than 1.7 billion offshore acres in the waters of the outer continental shelf.[5] Companies that develop and produce oil and gas from these federal lands and waters do so over a specified period of time under leases obtained from and administered by agencies of Interior—the Bureau of Land Management (BLM) for onshore leases and the Bureau of Ocean Energy Management (BOEM) for offshore leases. Interior's Office of Natural Resources Revenue (ONRR) is responsible for collecting revenues from onshore and offshore leases.

Interior acts on behalf of the American people to manage the federal oil and gas system to ensure a fair return to the public for the development of oil and gas resources. For offshore resources, one of the stated purposes of the Outer Continental Shelf Lands Act, as amended—which governs the management of oil and gas resources—is to ensure the public "a fair and equitable return" on the resources on the outer continental shelf. The law directs the Secretary of the Interior to conduct leasing activities to assure receipt of fair market value for the lands leased and the rights conveyed by the federal government. Further, for onshore resources, Interior relies on the competitive leasing process required by the Mineral Leasing Act, as amended, to ensure fair market value for onshore oil and gas resources. Broadly, we refer to the government collecting an appropriate share of revenue from leasing and production activities on federal lands and waters as ensuring a fair return. Because the revenues these leases generate depend, in part, on the amounts of oil and natural gas that companies produce from them, the federal government has sought to design fiscal systems that balance the goal of providing a fair return with sufficient financial incentives for companies to commit resources to exploring, developing, and producing oil and gas from their leases.

You asked us to review Interior's collection of oil and gas revenues as part of our ongoing efforts to support congressional oversight of GAO's high-risk areas. This report examines the steps Interior has taken to ensure that the public receives a fair return on federal oil and gas resources since 2007. To conduct this work, we reviewed applicable law, regulations, and guidance that govern Interior's management of oil and gas resources including the Outer Continental Shelf Lands Act, the Mineral Leasing Act, and the Federal

Onshore Oil and Gas Leasing Reform Act; examined our prior reports and Interior leasing program policies and documents; and interviewed Interior officials from BOEM, BLM, ONRR, and the Solicitor's Office. We also conducted a high-level review of the Interior-contracted studies of the fiscal system in order to summarize the studies' purpose and goals. We also compared actions that Interior took to standards for internal control in the federal government.[6]

We conducted this performance audit from October 2012 to December 2013 in accordance with generally accepted government auditing standards. Those standards require that we plan and perform the audit to obtain sufficient, appropriate evidence to provide a reasonable basis for our findings and conclusions based on our audit objectives. We believe that the evidence obtained provides a reasonable basis for our findings and conclusions based on our audit objectives.

BACKGROUND

This section describes (1) the federal oil and gas fiscal system, (2) leasing processes, and (3) the history of oil and gas management challenges. Created by Congress in 1849, Interior oversees the nation's publicly owned natural resources including parks, wildlife habitat, and crude oil and natural gas resources on millions of acres onshore and offshore in the waters of the outer continental shelf. With regard to oil and gas in particular, Interior leases federal lands and waters (also referred to as submerged lands), issues permits for oil and gas drilling, and is responsible for ensuring that the federal government receives payment from the private companies that extract oil and gas from federal leases.

The Federal Oil and Gas Fiscal System

The oil and gas fiscal system defines the applicable payments to the government from companies that lease federal lands and waters for oil and gas development. These payments include royalties, rents, and other payments—items generally specified within the lease terms. The revenues collected by the federal government on oil and gas development are shared with states, as directed by statute, and the remaining funds are deposited in the U.S. Treasury. In addition to the collection of these payments by Interior, the federal government assesses taxes on the profits companies earn on the sale of oil and gas produced from federal leases. Under the oil and gas fiscal system, companies bid on leases that Interior makes available. Interior awards the lease to the highest bidder generally based on a lump-sum payment called a bonus bid that is due when the lease is issued. The lease is a contract and conveys the rights to explore for and produce the oil and gas in a specified area to a company that holds the lease. The company is then subject to the payment of rental rates until production begins and then to payment of royalties on any oil and gas that is eventually produced on the lease. The royalty rate is a percentage of the value of production, and the royalty owed is the volume of production times the unit value of production times the royalty rate. The federal government receives royalty payments once production starts. In fiscal year 2012, the $9.7 billion in oil and gas revenue collected included royalties (about $8.5 billion or 87 percent), bonus bids (about $947 million or 10 percent),[7] and rental fees (about $272 million or 3 percent).

Currently Interior has the authority to change certain lease terms—such as the duration of the lease, royalty rates, and rental fees—within the overall oil and gas fiscal system. For new offshore leases, Interior is allowed by statute to change the lease terms for the bonus bid structure, rent, and royalty rates.[8] For new onshore leases, Interior is generally allowed by statute to change these same lease terms but with certain limits on flexibility. For onshore leases, Interior's regulations—issued in the 1980s—currently establish a royalty rate of 12.5 percent.[9] As such, changes to onshore royalty rates would require Interior to revise its regulations. With regard to taxes on corporate profits, only Congress may change the tax components of the oil and gas fiscal system as Interior does not have the authority to do so.

For both offshore and onshore leases, ONRR collects revenue from companies for the royalties, rents, bonuses, and other revenues generated throughout the leasing process. In this regard, ONRR has the responsibility to ensure that these revenues are accurately reported and paid in compliance with laws, regulations, and lease terms. ONRR establishes the regulations for how oil and gas are valued for royalty purposes, which affects the royalties that companies pay.

Leasing Processes

Interior's processes for issuing federal leases vary depending on whether they are offshore or onshore.

Offshore Leases

For offshore leases, management of oil and gas resources is primarily governed by the Outer Continental Shelf Lands Act, which sets forth procedures for leasing, exploration, and development and production of those resources. BOEM is the bureau within Interior responsible for implementing these requirements of the act related to preparing the leasing program. The act calls for the preparation of an oil and gas leasing program designed to best meet the nation's energy needs while also taking into account a range of principles and considerations specified by the act. Specifically, the act provides that "[m]anagement of the outer Continental Shelf shall be conducted in a manner which considers economic, social, and environmental values of the renewable and nonrenewable resources contained in the outer Continental Shelf, and the potential impact of oil and gas exploration on other resource values of the outer Continental Shelf and the marine, coastal, and human environments."[10] Furthermore, the act provides that the outer continental shelf is a "vital national resource reserve held by the federal government for the public, which should be made available for expeditious and orderly development, subject to environmental safeguards, in a manner which is consistent with the maintenance of competition and other national needs."[11] The act grants the Secretary the authority to issue leases and states that "leasing activities shall be conducted to assure receipt of fair market value for the lands leased and the rights conveyed by the Federal Government."[12]

The Outer Continental Shelf Lands Act requires the Secretary of the Interior to prepare an oil and gas leasing program that consists of a 5- year schedule of proposed lease sales that shows the size, timing, and location of leasing activity as precisely as possible. Every 5 years, Interior selects the areas it will offer for leasing and establishes a schedule for individual lease sales. These leases are offered for competitive bidding, and all eligible companies are invited to submit written sealed bids for the lease and rights to explore, develop, and produce oil and gas resources on these leases. These rights last for a set period of time, referred to as the initial period of the lease,[13] which varies depending on the water depth.[14] Interior estimates the fair market value of each lease, and the minimally acceptable bid is derived from this estimate. The bidder that submits the highest bonus bid that meets or exceeds Interior's minimum bid is awarded the lease. If a high bid does not satisfy any of the required conditions, the bid is rejected. In the event that no bid is received or no bids equal or exceed the minimum bid, Interior may choose to withdraw the lease—possibly offering it again at a future date.

Onshore Leases

For onshore leases, BLM's current leasing process was established in the Mineral Leasing Act of 1920, as amended, and the Mineral Leasing Act for Acquired Lands of 1947, as amended. Interior relies on the competitive leasing process required by the Mineral Leasing Act to ensure fair market value for onshore oil and gas resources. In addition, the Federal Land Policy and Management Act, though not specific to federal oil and gas resources, calls for the management of public lands in a manner that protects historical and environmental resources, provides for recreational and other uses, and ensures "fair market value" is received for their use and resources. BLM offers parcels of land nominated by industry and the public, as well as some it identifies. As with offshore leases, Interior initially offers onshore leases through a competitive bidding process; however, bonus bids are received in an oral auction rather than in a sealed written form, and Interior does not evaluate bid adequacy on a parcel-by-parcel basis. Instead, by law, it requires a uniform national minimum acceptable bid of $2 per acre that the Secretary has the authority to raise. If Interior receives any bids on an offered lease, the lease is awarded to the highest bidder. All onshore leases that do not receive any bids in the initial offer must be offered noncompetitively the day after and remain available for noncompetitive leasing for a period of 2 years after the competitive lease sale. Any of these available leases may be acquired noncompetitively on a first-come, first-served basis for the minimum acceptable bid. About 40 percent of existing BLM oil and gas leases were issued as noncompetitive leases.[15] For all competitively issued leases, the winning bidder must pay Interior the full amount of the bonus bid to become the lessee. The lessee then pays a fixed amount of rent each year until the lease begins producing or the lease terminates, expires, is cancelled, suspended, or relinquished.

History of Oil and Gas Management Challenges

In the 1970s and early 1980s, Interior's management of the oil and gas revenue collection system faced criticism by us and Interior's Office of the Inspector General. Interior's Inspector General issued five reports critical of the program between 1969 and 1977 and, in 1981, we reported that Interior was not collecting potentially hundreds of millions in royalties due from federal oil and gas leases.[16] In response, in 1981, the Secretary of the Interior established the Commission on Fiscal Accountability of the Nation's Energy

Resources, better known as the Linowes Commission named for the chairman of the commission, to investigate allegations of irregularities in royalty payments, among other issues. The Linowes Commission raised a number of serious concerns and in its report stated that "management of royalties for the nation's energy resources has been a failure for more than 20 years. Because the Federal government has not adequately managed this multibillion dollar enterprise, the oil and gas industry is not paying all the royalties it rightly owes."[17] The report cited a range of problems, including the failure to verify data reported by companies and late payments and underpayments, and concluded that, "[i]n short, the industry is essentially on an honor system." Among its 60 recommendations for improving the fiscal accountability of onshore and offshore resources, the commission called for raising onshore royalty rates to "appropriate levels." Specifically, the commission recommended that the onshore royalty rate for oil and gas be raised from 12.5 percent to 16.67 percent generally for new and renegotiated leases consistent with offshore royalty rates of 16.67 percent in place at that time.

Following the work of the commission, Interior and Congress took several actions aimed at improving management of revenue collection. In particular, the Secretary of the Interior, by secretarial order, reorganized the task of administering revenue collection under a new bureau; specifically, the Minerals Management Service (MMS) was created within Interior, in part, from the division of the U.S. Geological Survey—which was originally tasked with administering revenue collection, among other duties—to improve management of federal leasing revenues.[18] In addition, Congress passed legislation aimed at improving the collection of revenue including the Federal Oil and Gas Royalty Management Act of 1982 and the Federal Oil and Gas Royalty Simplification and Fairness Act of 1996.

In 2007, Interior's Subcommittee on Royalty Management—a subcommittee of the Royalty Policy Committee chartered to provide advice on royalty management issues and other mineral-related policies to the Secretary and other departmental officials responsible for managing mineral leasing activities—reported that a number of aspects of royalty management activities required prompt and, in some cases, significant management attention to ensure public confidence.[19] In particular, the report included over 100 recommendations to improve Interior's management of oil and gas resources, including those aimed at revising its valuation regulations and guidelines that govern the valuation of oil and gas resources for royalty purposes. According to Interior documentation, as of August 2012, 15 recommendations remain open.

In our May 2007 report, we found that, based on results of a number of studies, the government receives one of the lowest government takes in the world.[20] In September 2008,[21] we found that the fiscal system needed comprehensive reassessment and that Interior did not routinely evaluate the federal oil and gas fiscal system. Interior disagreed with recommendations in the draft report that it perform a comprehensive review of the fiscal system using an independent panel and adopt policies and procedures to keep abreast of important changes in the oil and gas market and in other countries' efforts to adjust their oil and gas management practices in light of these changes. Thus, in the final report, we suggested that Congress should consider directing the Secretary of the Interior to (1) convene an independent panel to perform a comprehensive review of the federal system for collecting oil and gas revenue and (2) establish procedures for periodically collecting data and information and conducting analyses to determine how the federal government take and the attractiveness for oil and gas investors in each federal oil and gas region compare to those of other resource owners and report this information to the Congress. Actions taken in response to this suggestion are discussed later in this report. In 2011, in part because of the challenges identified in our past work, we added Interior's management of federal oil and gas resources to our list of programs at high risk of fraud, waste, abuse, and mismanagement.[22] In the 2013 update of the high-risk list, we found that some progress had been made related to Interior's management of federal oil and gas resources and narrowed the federal oil and gas high-risk area to focus, in part, on the remaining issues related to revenue collection and ensuring that the public is getting an appropriate share of oil and gas revenues.[23]

Interior Has Taken Some Steps to Help Ensure a Fair Return but Does Not Have Procedures for Periodically Conducting Assessments of the Fiscal System

Interior has taken some steps to help ensure a fair return on federal oil and gas resources since our 2007 report, including: (1) changing offshore lease terms, while considering but not making changes to onshore lease terms; (2) contracting for studies of various aspects of the fiscal system; and (3) examining potential regulatory changes that could simplify royalty payments and collections. However, Interior does not have documented procedures in place for determining when to periodically conduct assessments of the fiscal system to ensure a fair return or for determining whether and how to make changes to lease terms for new offshore leases.

Interior Has Taken Some Steps Aimed at Ensuring a Fair Return

Interior has taken some steps aimed at ensuring a fair return, including changing offshore lease terms—such as increasing royalty rates, minimum bids, and rental rates—but onshore lease terms have not changed in recent years though onshore and offshore leasing programs are subject to many of the same market conditions. For example, while onshore royalty rates have remained at 12.5 percent, certain offshore royalty rates began increasing in 2007 to the current offshore royalty rate of 18.75 percent. Figures 1 and 2 depict changes to offshore and onshore royalty rates along with oil and gas price fluctuations, respectively, from January 2000 through July 2013. In addition, Interior has contracted for studies of various aspects of its fiscal system, including an assessment of how the federal fiscal system compared with the systems of other oil and gas resource owners (including owners in other countries); an analysis of policies that affect the pace of leasing in the Gulf of Mexico; and an analysis of the benefits, costs, and economic impacts of raising onshore royalty rates. Interior is also examining potential regulatory changes that could simplify royalty payments and collections.

Interior Changed Offshore Lease Terms

In recent years, Interior changed some offshore lease terms in an effort to ensure a fair return on oil and gas resources. Since 2007, Interior increased offshore lease terms including royalty rates, rental rates, and the minimum bid for certain offshore leases as follows:

- *Increased royalty rates.* From 2007 through 2008, Interior increased royalty rates for new leases by 50 percent. In 2007, Interior increased the royalty rate for new Gulf of Mexico leases from 12.5 percent to 16.67 percent for new leases in water depths greater than 400 meters. In 2008, Interior increased the rate again for all Gulf of Mexico leases to 18.75 percent. As of August 2013, all Gulf of Mexico royalty rates for new leases are 18.75 percent.[24] According to Interior officials and documents, incremental increases in royalty rates were instituted in response to a variety of factors including (1) increased oil and gas prices; (2) perceived improvements in exploration and production technologies, especially in deep water; and (3) the competitive market for offshore leases. Interior estimated that the royalty rate increase from 16.67 percent to 18.75 percent would result in a net increase in the total Gulf of Mexico federal revenues from bonuses, rents, and royalties from new leases of $4.3 billion, a 5 percent increase from $87.4 to $91.7 billion over 30 years.[25] After this 2008 royalty rate increase, Interior documents stated that demand remained strong for newly offered leases in the Gulf of Mexico and that Interior observed strong bidding interest in the three subsequent lease sales.[26]

- *Escalating and increased rental rates*. Interior established escalating rental rates—rates that increase over the duration of the lease—to encourage faster exploration and development of leases, or earlier relinquishment when exploration is unlikely to be undertaken by the lessee. Specifically, in 2007, Interior implemented escalating rental rates for leases offered in less than 400 meters of water—and in 2009, for leases offered in at least 400 meters of water.[27] Also, in 2009, Interior increased rental rates for new Gulf of Mexico leases in all water depths. Interior estimated that the increased rental rates and escalating rent rates in water depths greater than 400 meters would result in five fewer lease tracts receiving bids but an increase in rental revenue of $57 million over the initial lease term for leases resulting from that sale. $27 million of this $57 million was attributed to the increase in base rental rates.[28] In addition, the increased rental rates did not appear to reduce the number of lease blocks to be explored, according to Interior documents.
- *Increased minimum bids*. In 2011, Interior increased the minimum bid for leases offered in at least 400 meters of water in the Gulf of Mexico to $100 per acre, up from $37.50 per acre.[29] According to Interior's *Proposed Final Outer Continental Shelf Oil & Gas Leasing Program 2012-2017*, the minimum bid was raised, in part, to account for increases in oil prices and to encourage optimal timing of leasing. Interior officials told us that a review of the minimum bid was initiated because the minimum bid had not been changed in some time. In addition, Interior analysis showed that a minimum bid of $100 per acre would be generally equivalent to the cost of the minimum bid in the past, going back to 1999, adjusted for differences in prices, costs, and royalty rates.

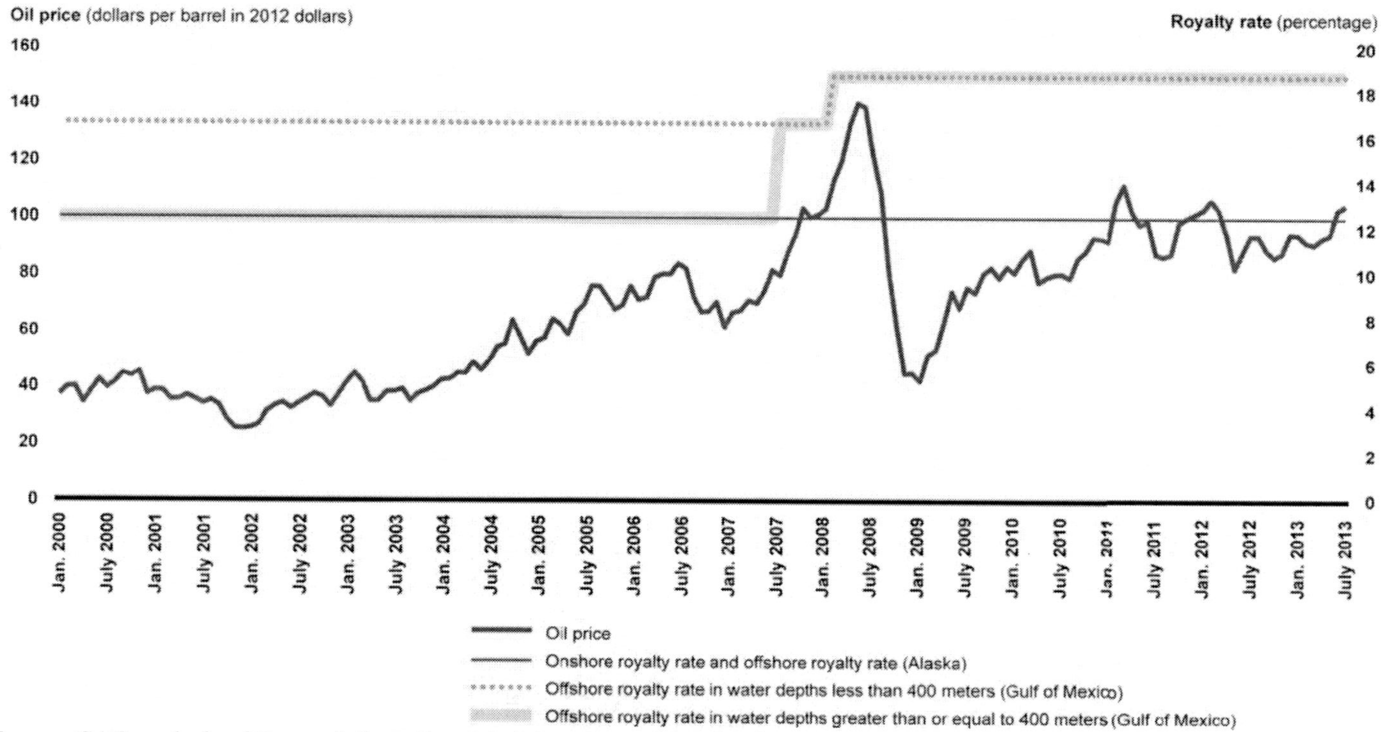

Source: GAO analysis of Energy Information Administration and Interior data.

Figure 1. Royalty Rates and Oil Prices, January 2000 through July 2013.

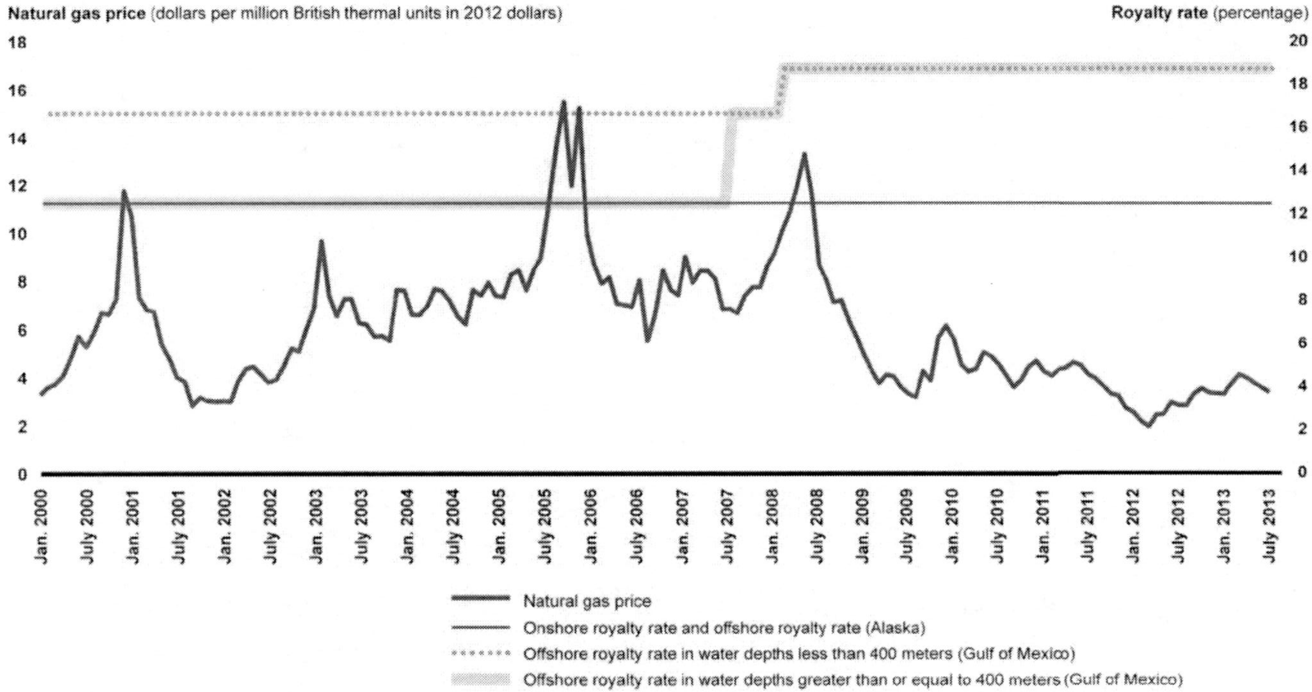

Source: GAO analysis of Energy Information Administration and Interior data.

Figure 2. Royalty Rates and Gas Prices, January 2000 through July 2013.

Table 1. History of Lease Terms for Gulf of Mexico Outer Continental Shelf Oil and Gas Leases 2005-2013

Lease sale	Sale date	Lease terms	Water depth (meters)					
			0-200	**200-400**	**400-800**	**800-1,600**	**1,600-2,000**	**>2,000**
196	8/17/2005	Rent ($/acre)	6.25	9.50	9.50	9.50	9.50	9.50
		Minimum bid ($/acre)	25.00	25.00	37.50	37.50	37.50	37.50
		Royalty rate	16.67%[a]	16.67%	12.50%[a]	12.50%[a]	12.50%[a]	12.50%[a]
198 and 200	3/15/2006 - 8/16/2006	Rent ($/acre)	6.25	9.50	9.50	9.50	9.50	9.50
		Minimum bid ($/acre)	25.00	25.00	37.50	37.50	37.50	37.50
		Royalty rate	16.67%[a]	16.67%[a]	12.50%[a]	12.50%[a]	12.50%[a]	12.50%[a]
204 and 205	8/22/2007 - 10/3/2007	Rent ($/acre)	**6.25**[b]	**9.5**[b]	9.50	9.50	9.50	9.50
		Minimum bid ($/acre)	25.00	25.00	37.50	37.50	37.50	37.50
		Royalty rate	16.67%[a]	16.67%[a]	**16.67%**[a]	**16.67%**[a]	**16.67%**[a]	**16.67%**[a]
224	3/19/2008	Rent ($/acre)				9.50	9.50	9.50
		Minimum bid ($/acre)				37.50	37.50	37.50
		Royalty rate				**18.75%**	**18.75%**	**18.75%**
206 and 207	3/19/2008 - 8/20/2008	Rent ($/acre)	6.25[b]	9.5[b]	9.50	9.50	9.50	9.50
		Minimum bid ($/acre)	25.00	25.00	37.50	37.50	37.50	37.50
		Royalty rate	**18.75%**[a]	**18.75%**[a]	**18.75%**[a]	18.75%[a]	18.75%[a]	18.75%[a]
208	3/18/2009	Rent ($/acre)	**7.00**[b]	**11.00**[b]	**11.00**[b]	**11.00**[b]	**11.00**[b]	**11.00**[b]
		Minimum bid ($/acre)	25.00	25.00	37.50	37.50	37.50	37.50
		Royalty rate	18.75%[a]	18.75%[a]	18.75%[a]	18.75%[a]	18.75%[a]	18.75%[a]
210 and 213	8/19/2009 - 3/17/2010	Rent ($/acre)	7.00[b]	11.00[b]	11.00[b]	11.00[b]	11.00[b]	11.00[b]
		Minimum bid ($/acre)	25.00	25.00	37.50	37.50	37.50	37.50
		Royalty rate	18.75%[a]	18.75%[a]	18.75%[a]	18.75%[a]	18.75%[a]	18.75%[a]

Lease sale	Sale date	Lease terms	Water depth (meters)					
			0-200	200-400	400-800	800-1,600	1,600-2,000	>2,000
218 to 229	12/14/2011 - 3/20/2013	Rent ($/acre)	7.00[b]	11.00[b]	11.00[b]	11.00[b]	11.00[b]	11.00[b]
		Minimum bid ($/acre)	25.00	25.00	**100.00**	**100.00**	**100.00**	**100.00**
		Royalty rate	18.75%[a]	18.75%[a]	18.75%	18.75%	18.75%	18.75%
233	8/28/2013	Rent ($/acre)	7.00[b]	11.00[b]	11.00[b]	11.00[b]	11.00[b]	11.00[b]
		Minimum bid ($/acre)	25.00	25.00	100.00	100.00	100.00	100.00
		Royalty rate	18.75%[a]	18.75%[a]	18.75%	18.75%	18.75%	18.75%

Source: GAO analysis of Interior data.

Note: One meter equals about 3.28 feet.

[a] Leases may be eligible for royalty relief. Certain leases include royalty relief provisions for shallow water deep gas, and other leases may be eligible to apply for shallow water deep gas royalty relief. Leases resulting from sales held after 2000 may be issued with certain royalty relief provisions, and all leases obtained after 2000 in water depths greater than 200 meters are also eligible to apply for royalty relief.

[b] Rent per acre is for years 1 through 5 of the lease; then the rental rate escalates in year 6 and, in some cases, increases again in subsequent years as well.

For details on the recent history of lease terms in the Gulf of Mexico, see table 1; changes in lease terms are highlighted in gray.

Interior has also taken actions to encourage the development of oil and gas resources, which reduces the time from when federal leases are issued and the federal government receives its share of revenue from them, in response to our October 2008 recommendation that the Secretary of the Interior develop a strategy to evaluate options to encourage faster development of its oil and gas leases.[30] Specifically, in 2010, Interior shortened lease terms by reducing the duration of the initial period for Gulf of Mexico leases in water depths of 400 to less than 800 meters from an 8-year initial period to a 5-year initial period. For water depths of 800 to less than 1,600 meters, it reduced leases from a 10-year initial period to a 7-year initial period. According to Interior documents, these lease terms can generally be extended if the lessee begins drilling a well during the initial period.

Interior Has Considered but Has Not Changed Onshore Lease Terms

For onshore resources, Interior has considered, but not made, changes to onshore lease terms in order to provide greater assurance that the public is getting a fair return on federal oil and gas resources. Interior acts on behalf of the American people to manage the federal oil and gas system to ensure a fair return to the public for the development of oil and gas resources. Interior officials told us that since 2009 the department has been considering increasing the onshore royalty rate—which is currently established in its regulations at 12.5 percent for both oil and gas.[31] According to the officials, several factors prompted efforts to consider changing the royalty rates, including our September 2008 report,[32] oil and gas prices, and Office of Management and Budget initiatives calling for increased revenue from onshore royalties.[33] Although both onshore and offshore leasing programs are subject to many of the same market conditions, Interior officials are currently unable to make timely adjustments to onshore royalty rates because BLM's regulations generally establish a set royalty rate of 12.5 percent. This limits the bureau's flexibility because making adjustments to that rate require going through the rulemaking process, and the process can take several years according to Interior officials. Specifically, officials said that the public notice and comment period required as part of the rulemaking process could take 1 to 2 years, and proposed rules must also undergo review by the Office of Management and Budget.

Interior officials told us that the department planned to publish a notice of proposed rulemaking in July 2012 to change BLM's regulations to set an onshore royalty rate of 18.75 percent for oil production on new federal competitive leases but leave the royalty rate for gas production unchanged at 12.5 percent. The planned regulatory revisions would have allowed the Secretary to review and revise royalty rates for new competitive leases as appropriate—similar to the authority that the Secretary has for revising offshore royalty rates. Officials told us that including the requirement for periodic review and revision of royalty rates would have given the Secretary greater flexibility to go forward with such reviews and revisions in the future.

Interior discontinued its efforts to pursue the revised regulations because, according to Interior officials, the department does not have enough information to determine how to adjust onshore royalty rates. Rather, Interior plans to ask the public to comment on whether and how royalty rates for new federal onshore competitive oil and gas leases should be revised to better ensure a fair return to the public. Specifically, Interior officials told us they plan to ask for comments on the types of royalty rate structures that should be considered, such as whether BLM should develop a uniform rate for all leases or different rates by region, state, geologic formation, or resource type. Furthermore, Interior officials told us they would also ask for comments on whether sliding scale royalty rates—or rates that vary with the price of the commodity—might be appropriate in specific circumstances. An Advance Notice Of Proposed Rulemaking is under development, but officials told us that higher priority rulemaking initiatives, such as regulations for hydraulic fracturing and revisions to its oil and gas measurement regulations, precede it and that limited resources constrain their ability to meet program demands. As a result of not successfully changing federal regulations to provide itself with the flexibility needed to make timely adjustments to onshore lease terms, Interior's ability to ensure that the public is receiving a fair return is limited. Moreover, Interior continues to offer onshore leases with lease terms—terms lasting the life of the lease—that have not been adjusted in response to changing market conditions, potentially foregoing a considerable amount of revenue. For example, in 2011, Interior estimated that onshore royalty rate changes could increase revenue collections by about $1.25 billion over 10 years.[34]

Interior Contracted for Studies of Various Aspects of the Fiscal System

Interior contracted for several studies—including a study of how the federal oil and gas fiscal system compared with fiscal systems of other resource owners—that reviewed various aspects of the federal oil and gas fiscal system since 2007. In our September 2008 report,[35] we found that

Interior collected a lower government take for oil and gas production in the deep water of the U.S. Gulf of Mexico than all but 11 of 104 oil and gas resource owners whose revenue collection systems were evaluated in a comprehensive industry study, which included other countries as well as some states.[36] We also found that Interior had not routinely evaluated the federal oil and gas fiscal system, monitored what other governments or resource owners were receiving for their resources, or evaluated and compared the attractiveness of federal lands and waters for investment with that of other regions.[37] In response to our 2008 findings, Interior contracted for a study—the 2011 *Comparative Assessment of the Federal Oil and Gas Fiscal System* study—that compared the federal oil and gas fiscal systems of selected federal oil and gas regions to that of other resource owners. In addition, Interior contracted for two other studies on the effect of different leasing and royalty rate policies on revenue, exploration, and production. See table 2 for a description of these studies.

Table 2. Interior-Contracted Studies of the Oil and Gas Fiscal System 2009-2013

Study	Description of analysis conducted
Comparative Assessment of the Federal Oil and Gas Fiscal System (October 2011)[a]	Examines 29 fiscal systems—including the current systems relating to federal offshore oil and gas resources in the Gulf of Mexico and onshore gas resources in Wyoming—and describes the impact of various lease term changes on other aspects of the system such as the system's stability and competitiveness, pace of leasing, and revenue. The study provided information on fiscal system components, such as royalty rates and taxes, for specific areas within the United States and other countries. Identifies four fiscal-related factors—government take, internal rate of return, profit- investment ratio, and progressivity—and constructs a hypothetical, composite index using these measures to compare fiscal systems. The report provides an assessment of how changes to the royalty rate could potentially affect industry interest in federal offerings.
Benefit-Cost and Economic Impact Analysis of Raising the Onshore Royalty Rates Associated with New Federal Oil and	Assesses the impacts of raising onshore royalty rates associated with new competitive leases and considers both fixed and sliding scale royalty rates. The study also addresses changes to the net economic benefits to the states and federal government, changes to the demand for federal leases and changes to production. Estimates the impact of royalty rate changes—including

Study	Description of analysis conducted
Gas Leasing (April 2011)[b]	sliding scale scenarios—on revenue and other parts of the revenue stream, such as bonus bids, and analyzes high and low price scenarios for oil and gas.
Policies to Affect the Pace of Leasing and Revenues in the Gulf of Mexico (August 2009/December 2010)[c]	Examines alternative leasing policies for outer continental shelf oil and gas resources in the Central and Western Gulf of Mexico. The study focuses on tracts to be leased in the Central and Western Gulf of Mexico planning areas over the 50-year period from 2010 – 2060. Considers several leasing policies and estimates the impact on exploration, production, and revenues. The study analyzes alternative leasing systems and describes the goals and criteria for assessing alternative leasing systems. The study models the various potential leasing systems and compares them with the status quo. Leasing systems' alternatives considered include slowing the pace of leasing, changing royalty rates, raising minimum bids, profit sharing, raising area rental payments, using different bidding systems, implementation of work commitments, and reducing the length of the primary lease period.

Source: GAO analysis of Interior-contracted studies.

[a] IHS Cambridge Energy Research Associates, *Comparative Assessment of the Federal Oil and GasFiscal System* (October 2011).

[b] Enegis, LLC, *Benefit-Cost and Economic Impact Analysis of Raising the Onshore Royalty Rates Associated with New Federal Oil and Gas Leasing* (April 2011). This report was contracted by BLM to help guide the decision-making process with regards to a proposed royalty rate rulemaking. The rulemaking is still under deliberation by the agency, and all materials related to this process are considered pre-decisional in nature by Interior.

[c] Economic Analysis, Inc. and Marine Policy Center, *Policies to Affect the Pace of Leasing and Revenues in the Gulf of Mexico* (August 2009/December 2010). Each version of this study evaluates the same policy alternatives under different assumptions about the basic conditions for future development in the Gulf of Mexico such as the ultimate resource size in the Gulf of Mexico and lessee actions related to potential future effective tax rates.

According to Interior officials, the study conducted in response to our 2008 findings—the 2011 *Comparative Assessment of the Federal Oil and Gas Fiscal System*—provided some useful information about the fiscal system such as how fiscal terms in the United States compared with other resource owners, but it has not directly led to any changes to the fiscal system or lease terms for new federal oil and gas leases. Similarly, Interior officials told us that the other two studies have not yet led to revisions to the fiscal system or lease terms for new offshore or onshore leases. Rather, according to officials, additional internal analyses and modeling, as well as consultation with stakeholders—including oil and gas companies and the public—will continue to primarily inform future changes to the fiscal system. Moreover, Interior did not document any internal discussions or analysis of the three studies' findings. As part of the 2011 *Comparative Assessment of the Federal Oil and Gas Fiscal System* study, officials said that they obtained a model that can be employed in the future to conduct comparative analyses but currently have no plans to update the model or the data inputs used by the model. Officials told us that the study's findings reassured them that their own internal assessment related to the competitiveness of the offshore fiscal system was appropriate. In addition, officials said that the study provided additional information—mainly raising the issue of whether an appropriate return was being received for onshore resources—but that the study was not adequate to determine next steps for onshore lease terms.

Interior Is Examining Potential Regulatory Changes that Could Simplify Royalty Payments and Collections

Interior is examining potential regulatory changes that could simplify royalty payments and collections. As we found in our past work,[38] complex valuation regulations can result in inaccurate royalty payments made by industry, and this could increase ONRR's costs to ensure accurate royalty payments because of the need for potentially detailed and time-consuming audits of records. In May 2011, Interior published an Advance Notice Of Proposed Rulemaking requesting comments to inform potential changes to regulations intended to simplify royalty payments and collections. In addition to our work, others have identified numerous shortcomings in ONRR's royalty collection programs, in part because of its valuation regulations' complex requirements for calculating the value of oil and gas and associated deductions and allowances for activities such as transportation.

In December 2007, Interior's Subcommittee on Royalty Management recommended that, by the end of fiscal year 2008, Interior publish proposed revisions to the gas valuation regulations to, among other goals, simplify the calculation of royalties and deductions for gas transportation and processing. Interior did not meet this time frame due to several factors including the complexity of oil and gas valuation, according to Interior officials. In May 2011, Interior published the Advance Notice Of Proposed Rulemaking in the *Federal Register* for the proposed rule, which according to Interior documents is intended, in part, to provide greater simplicity, certainty, clarity, and consistency in production valuation; decrease ONRR's costs to ensure compliance; decrease industry's compliance costs; and provide more certainty to ONRR and industry that companies pay every dollar due to the government.[39] According to ONRR officials, the proposed regulations were undergoing internal review as of September 2013 and are expected to be published in the *Federal Register* in 2014.

Interior Does Not Have Documented Procedures for Conducting Periodic Assessments of the Fiscal System or for Supporting Potential Changes to New Lease Terms

Interior does not have documented procedures in place for determining (1) when to conduct periodic assessments of the overall fiscal system or (2) whether and how to make changes to lease terms for new offshore leases.

Interior Does Not Have Documented Procedures for Determining When to Conduct Periodic Assessments of the Overall Fiscal System

Interior does not have documented procedures in place for determining when to periodically conduct assessments of the overall fiscal system as a whole. Although Interior recently contracted for such an assessment, it was the first in well over 25 years. Without documented procedures, Interior cannot ensure that it will consistently conduct such assessments in the future, and without periodically conducting such assessments, Interior cannot know whether there is a proper balance between the attractiveness of federal leases for investment and appropriate returns for federal oil and gas resources, limiting Interior's ability to ensure a fair return on federal oil and gas resources. Internal control standards in the federal government call for agencies to clearly document internal controls and the documentation is "to appear in management directives, administrative policies, or operating manuals."[40] Documented procedures of when Interior is to conduct such assessments—whether within specific time frames or the occurrence of certain market or industry changes— could help provide the department with reasonable assurance that its staff knows when to conduct assessments of the overall fiscal system to help ensure those reviews are conducted systematically and consistently.

In our September 2008 report, we found that the last time Interior conducted a comprehensive assessment of the federal oil and gas fiscal system was over 25 years ago. Additionally, we reported that, without routinely evaluating the federal oil and gas system as a whole, including monitoring what other resource owners worldwide are receiving for their energy resources or evaluating and comparing the attractiveness of the United States for oil and gas investment with that of other oil and gas regions, Interior cannot provide reasonable assurance that the public is getting an appropriate share of revenues.[41] As mentioned previously, in response to our 2008 findings, Interior contracted for a study that compared the federal oil and gas fiscal system to that of other resource owners. As part of this study, officials said that they obtained a model that can be employed in the future to conduct comparative analyses; however, there are currently no plans to update the model or the data inputs used by the model. Interior officials told us that this type of comprehensive assessment would only be undertaken if fundamental shifts in the market occurred. According to officials, however, Interior does not have procedures or criteria in place for determining when such an assessment of the fiscal system should take place or what changes in the market or industry would signal that such a study should be done. Without procedures for determining when to conduct periodic assessments of the fiscal system as a whole, Interior cannot be reasonably assured that it will consistently conduct such assessments in the future, limiting its ability to be confident that the system is ensuring a fair return on federal oil and gas resources. According to the Office of Management and Budget, rigorous program evaluations can help determine whether government programs are achieving their intended outcomes to the extent possible.[42]

Moreover, Interior's oversight of federal land and waters is subject to the federal government's multiple, diverse objectives—fair return, protection of historical and environmental resources, and expeditious and orderly development, among other goals. Thus, Interior is confronted with evaluating these objectives in light of a complex set of factors—including market prices and how development opportunities in the United States compare with those of other resource owners. By having documented procedures, the department could help ensure that its evaluations take all of these factors into consideration. Further, these factors may change over time as the market for oil and gas changes, the technologies used to explore and produce oil and gas change, or as the broader economic climate changes, making it even more important that Interior has documented procedures for conducting periodic assessments of the federal oil and gas fiscal system.

In addition, Interior has conducted some analyses to support changes to offshore lease terms in advance of offshore lease sales, which typically occur a few times a year—but the analyses conducted to support these changes are not a substitute for periodically assessing the oil and gas fiscal system as a whole. Since 2007, Interior has conducted some analyses for offshore lease sales in support of changes to royalty rates, rental rates, and the minimum bid. Based on our review of Interior documents from several lease sales from 2007 to 2011, we found that the analyses the department conducted to support proposed changes to offshore lease terms generally involved estimating the impacts of a proposed change on revenue, bidding activity, and potential oil and gas production. In addition, Interior's documentation shows that the department took into consideration technological and market conditions; policy goals, such as promoting development or enhancing revenues; and administrative benefits, such as making lease terms consistent across water depths. However, these analyses did not include an evaluation of what other resource owners worldwide are receiving for their energy resources or a comparison of the attractiveness of the United States for oil and gas investment with that of other oil and gas resource owners. In our September 2008 report, we suggested that Congress should consider directing the Secretary of the Interior to establish procedures for periodically collecting data and information and conducting analyses to determine how the federal government take and the attractiveness for oil and gas investors in each federal oil and gas region compare with those of other resource owners and report this information to Congress. Moreover, Interior has not conducted similar analysis for onshore lease sales. Interior officials explained that the mechanism for determining the value of onshore resources is directed by statute to be market-driven.[43] Specifically, officials stated that fair market value for onshore lease sales is determined through the oral competitive bidding process required by the Mineral Leasing Act, rather than an evaluation of the geology and potential value of the oil and gas resource.[44] Because the leasing process is established by statute, Interior officials told us that it has not recently examined whether alternative leasing systems might be more effective in ensuring fair market value.

Interior Does Not Have Documented Procedures for Determining Whether and How to Make Changes to New Offshore Lease Terms

Interior does not have documented procedures for determining whether and how to make changes to new offshore lease terms—which are specified a few times per year ahead of each lease sale—consistent with federal standards for internal control. Without documented procedures for determining whether and how to make changes to new offshore lease terms, Interior is at risk of making inconsistent determinations about lease terms. Such inconsistent determinations would undermine its credibility and its ability to better ensure a fair return on oil and gas resources. Officials told us Interior does not have documented procedures or criteria for determining whether and how to make changes to offshore lease terms. Rather Interior officials said that they follow an informal process and establish offshore lease terms for each sale but that they do not have the time or resources to evaluate each lease term prior to each lease sale. However, based on our review of Interior documents, the analyses the department conducted to support proposed changes to offshore lease terms were inconsistent in the array of conditions and factors the department considered, and the level of analysis conducted in support of decisions to change lease terms varied and was not consistently or clearly documented. For example, as mentioned previously, escalating rental rates were implemented in two different sales—first, in 2007, in water depths less than 400 meters and then, in 2009, in water depths greater than 400 meters. The rationale provided for the first change was the policy goal to expedite drilling and compensate Interior, while the rationale in 2009 included specific estimates of the effects of an escalating rental rate on potential revenue and bidding, as well as consideration of market conditions. While both justifications may be warranted, because Interior does not have documented procedures specifying whether and how to support changes to its lease terms, Interior's approach to revising its lease terms appears to be inconsistent.

In addition, our review of documents supporting two separate royalty rate changes in 2007 and 2008—the first in 25 years—found that Interior did not consistently document the justifications and analysis supporting the increases. Specifically, Interior documents for the 2008 royalty rate increase cite reasons similar to the 2007 royalty rate increase—generally significant changes in market conditions—but because the second increase took place less than a year after the implementation of the first increase, it is unclear what significant changes in market conditions occurred to prompt the consideration of the second increase in royalty rates. Internal control standards in the federal government call for agencies to clearly document transactions and other significant events and that documentation should be readily available for examination.[45] While both royalty rate increases may have been warranted, clear documentation of the justifications and analysis supporting royalty rate increases would make Interior's decisions to change the royalty rates transparent and could inform future decision making related to changing rates. Such transparency can be particularly helpful in the event that key staff retire or leave federal service. Documentation of internal discussions that took place prior to the second royalty rate increase show that prior to being able to assess the impacts of increasing the royalty rate from 12.5 percent to 16.67 percent in 2007, Interior was considering an additional royalty rate increase. In addition, Interior documents show Interior program officials' concerns about an additional increase in royalty rates; specifically, officials urged the need to analyze the impact of the first increase, and they also noted potential negative effects of an increase including delaying investment and production in certain areas of the Gulf of Mexico. By having documented procedures for determining whether and how to make future changes to offshore leasing terms, Interior could increase its consistency and thus enhance its credibility in the conditions and factors the department considered and the level of analysis conducted.

Conclusion

Interior has taken several steps intended to help ensure that the public receives a fair return on oil and gas produced from federal leases.

However, even with these recent steps, it is not clear that Interior's efforts, by themselves, provide long-lasting assurance that federal resources will provide a fair return. This is especially true in light of the absence of documented procedures for Interior to determine when it will periodically conduct assessments of the overall federal oil and gas fiscal system and whether and how to make changes to new offshore lease terms, as well as Interior's limited flexibility to make changes to new onshore lease terms. Ensuring that the federal government is obtaining fair return for the resources it manages on behalf of its citizens is especially important as the country faces ongoing fiscal challenges.

Although leasing programs for both onshore and offshore areas are subject to many of the same market conditions, and Interior has increased offshore royalty rates, officials overseeing onshore leasing are currently unable to make timely adjustments to onshore royalty rates because, in general, BLM's regulations fix the rate at 12.5 percent, potentially limiting Interior's ability to ensure that the public is receiving a fair return and potentially resulting in foregone revenue. In particular, while Interior has changed offshore lease terms several times over the past few years in response to changes in market conditions—many of which also affect onshore areas—to better ensure a fair return, Interior has not successfully changed BLM's regulations to provide itself with the flexibility needed to change onshore lease terms in a timely manner despite considering increasing the onshore royalty rate since 2009. As a result, Interior continues to offer onshore leases with lease terms—terms lasting the life of the lease—that have not been adjusted in response to changing market conditions, potentially foregoing a considerable amount of revenue.

Key among Interior's efforts to ensure a fair return, to address a GAO recommendation, Interior completed an assessment—the *Comparative Assessment of the Federal Oil and Gas Fiscal System*—which examined how the fiscal system of selected federal oil and gas regions compared with fiscal systems of other resources owners. Interior officials told us, however, that it has no plans to update the assessment, increasing the risk that this progress may be fleeting and that, as we found in 2008, it could be years before another assessment is completed during which time there could be significant changes in market conditions. Furthermore, without documented procedures in place for conducting periodic assessments of the fiscal system—such as when such an assessment of the fiscal system should take place or what changes in the market or industry would signal that such a study should be done, Interior cannot know whether there is a proper balance between the attractiveness of federal leases for investment and appropriate returns for federal oil and gas resources, limiting Interior's ability to ensure a fair return.

Finally, while Interior has made changes to its offshore lease terms, for example increasing royalty rates in some instances from 12.5 percent to 18.75 percent, Interior does not have documented procedures for determining whether and how to make changes to lease terms for new offshore leases. Without such documented procedures, Interior's rationale is not transparent, and it is at risk of making inconsistent determinations about lease terms. Such inconsistency would undermine its credibility and ability to better ensure a fair return on oil and gas resources. Additionally, Interior has not clearly documented the justifications and analysis supporting changes to lease terms, including royalty rate increases. As a result, the department's decisions to change lease terms are not transparent and, without documentation of these decisions, its future efforts to change rates may be impeded.

RECOMMENDATIONS FOR EXECUTIVE ACTION

To better ensure that the government receives a fair return on its oil and gas resources, we recommend that the Secretary of the Interior take the following three actions:

- Take steps, within existing authority, to revise BLM's regulations to provide for flexibility to the bureau to make changes to onshore royalty rates, similar to that which is already available for offshore leases, to enhance Interior's ability to make timely adjustments to the terms for federal onshore leases.
- Establish documented procedures for determining when to conduct periodic assessments of the overall fiscal system. Such procedures should identify generally when such an assessment should be done or what changes in the market or industry would signal that such an assessment should be done. Additionally, the assessment should include determining how the government's share of revenue from the federal oil and gas fiscal system and the attractiveness for oil and gas investors in each federal oil and gas region compare with those of other resource owners.
- Establish documented procedures for determining whether and how to adjust lease terms for new offshore leases, including documenting the justification and analysis supporting any adjustments.

AGENCY COMMENTS

We provided a draft of this report to Interior for review and comment. Interior generally agreed with our findings and concurred with our recommendations.

Sincerely yours,
Frank Rusco
Director,
Natural Resources and Environment

End Notes

[1] *Fiscal Accountability of the Nation's Energy Resources* (January 1982).

[2] GAO, *Oil and Gas Royalties: A Comparison of the Share of Revenue Received from Oil and Gas Production by the Federal Government and Other Resource Owners,* GAO-07-676R (Washington, D.C.: May 1, 2007). The government take, which accrues to any government, is largely determined by the government's oil and gas fiscal system. In the United States, this fiscal system consists of both terms specified in the lease, such as the royalty rate and rent, as well as the corporate taxes paid by the company on profits from the sale of oil and gas produced from federal leases.

[3] GAO, *Oil and Gas Royalties: The Federal System for Collecting Oil and Gas Revenues Needs Comprehensive Reassessment*, GAO-08-691 (Washington, D.C.: Sept. 3, 2008). The status of these actions is discussed later in this report.

[4] GAO, *High-Risk Series: An Update,* GAO-11-278 (Washington, D.C.: Feb. 16, 2011).

[5] The outer continental shelf (submerged lands) is outside the territorial jurisdiction of all 50 states but within U.S. jurisdiction and control and consists of submerged federal lands, generally extending seaward between 3 and 200 nautical miles off the U.S. coastline.

[6] GAO, *Standards for Internal Control in the Federal Government*, GAO/AIMD-00-21.3.1 (Washington, D.C.: November 1999).

[7] Lease sale 222 was held in June 2012 and resulted in bonus revenues of $1.68 billion; however, since certain bid review procedures were not completed by the end of the fiscal year, not all of these bonuses are included in the revenue amount for bonus bids.

[8] 43 U.S.C. §1337 provides that Interior can make changes so long as there is only one bid variable or "flexible" term. The bid variable or "flexible" term does not fluctuate over the life of the lease but is the term on which bidders compete for the award through the level of the bid made.

[9] By statute, the Secretary of the Interior must first offer parcels at competitive lease sales and may only issue noncompetitive leases after the department has offered the lands competitively at an oral auction and not received a bid. For onshore leases issued noncompetitively, the Mineral Leasing Act sets a 12.5 percent royalty rate by statute.

[10] 43 U.S.C. §1344(a)(1).

[11] 43 U.S.C. §1332(3).

[12] 43 U.S.C. §1344(a)(4).

[13] If a discovery is made within the initial term of the lease, the lease is extended for as long as oil and/or natural gas is produced in paying quantities or approved drilling operations are conducted. The term of the lease may also be extended if a suspension of production or suspension of operations has been granted or directed.

[14] In the Gulf of Mexico, in a recent notice of sale in 2012, BOEM offered leases with an initial term of 5 years extended to 8 years if drilling begins during the initial 5-year period targeting hydrocarbons below a depth of at least 25,000 feet subsea for leases in less than 400 meters of water. For leases in 400 to 800 meters of water, the initial term was 5 years extended to 8 years if drilling begins during the initial 5-year period. For leases in 800 to 1,600 meters of water, the initial period was 7 years extended to 10 years if drilling begins during the initial 7-year period. For leases in over 1,600 meters of water, the initial period was 10 years.

[15] BLM, *Public Land Statistics 2012*, Volume 197 (June 2013). As of September 30, 2012, BLM reported 27,747 competitive oil and gas leases covering 22.2 million acres and 18,411 noncompetitive oil and gas leases covering 14.8 million acres.

[16] GAO, *Oil and Gas Royalty Collections: Longstanding Problems Costing Millions,* GAO/AFMD-82-6 (Washington, D.C.: Oct. 29, 1981). In our October 1981 report, we recommended that Interior, as part of its efforts to develop a new royalty accounting system, should (1) monitor the development of the new system and (2) include as part of these redesign efforts a plan for, among other actions, monitoring and reconciling records, inspecting leases, and verifying production and sales data. Interior did not take action to implement either of these recommendations in part because it had initiated a new review of its revenue collection system that covered many of the same issues we identified.

[17] *Fiscal Accountability of the Nation's Energy Resources* (January 1982).

[18] In 2010, following the explosion of the *Deepwater Horizon* drilling rig in the Gulf of Mexico, Interior announced that it was going to reorganize its offshore oversight and revenue collection functions. Specifically, Interior eventually restructured MMS into three separate bureaus—BOEM, responsible for offshore leasing and resource management; the Bureau of Safety and Environmental Enforcement, responsible for issuing oil and natural gas drilling

permits, environmental safety and regulation, and conducting inspections for offshore leases; and ONRR, responsible for revenue collection for both offshore and onshore leases.

[19] Subcommittee on Royalty Management, Royalty Policy Committee, *Report to the Royalty Policy Committee: Mineral Revenue Collection from Federal and Indian Lands and the Outer Continental Shelf* (Washington, D.C.: 2007).

[20] GAO-07-676R.

[21] GAO-08-691.

[22] GAO-11-278.

[23] GAO, *High-Risk Series: An Update,* GAO-13-283 (Washington, D.C.: Feb. 14, 2013).

[24] Although the outer continental shelf leases for the Gulf of Mexico have increased royalty rates, the outer continental shelf leases for Alaska have remained at a 12.5 percent royalty rate for about 30 years.

[25] The revenue estimates are nominal dollars unadjusted for inflation.

[26] Interior's analysis included estimates for increasing royalty rates beyond 18.75 percent. Specifically, it estimated that royalty rate increases from 18.75 to 21.875 percent would cause production losses of 2 to 6 percent with royalty revenue increases of 11 to 17 percent. According to the analysis, the effect of increased royalty rates, depending on the size of the change, would be less production, but with the potential for higher revenues from royalties in the future. Interior found that a large increase in the royalty rate could curtail expected returns to lessees to such a large extent that it might unduly reduce leasing and future production by proportions greater than suggested in its analysis. Much higher royalty rates could also curtail production from new leases in the future as production declines in the later phases of a lease's productive life.

[27] Under this change, the prevailing rental rates for new leases in water depths of less than 200 meters would be $7/acre for the first 5 years with increases to $14/acre in year 6, $21/acre in year 7, and $28/acre in year 8. For new leases in water depths from 200 to 400 meters, rental rates increase from $11/acre to $22/acre in year 6, $33/acre in year 7, and $44/acre in year 8. For new leases in water depths from 400 to 800 meters, rental rates increase from $11/acre to $16/acre in years 6 through 8. For new leases in water depths greater than 800 meters, rental rates increase from $11/acre to $16/acre in years 6 through 10.

[28] The revenue estimates are in nominal dollars unadjusted for inflation for leases resulting from sale 208 and the revenue estimates for the increase in base rental rates include estimates for the initial lease term.

[29] The Gulf of Mexico minimum bid remains at $25 per acre in water depths of less than 400 meters. The most recent minimum bids in Alaska were $25 per hectare (about $10 per acre) in the Chukchi Sea, Cook Inlet, and in Zone B (deeper water areas) of the Beaufort Sea; and $37.50 per hectare (about $15 per acre) in Zone A (near shore areas) of the Beaufort Sea.

[30] GAO, *Oil and Gas Leasing: Interior Could Do More to Encourage Diligent Development*, GAO-09-74 (Washington, D.C.: Oct. 3, 2008).

[31] For onshore leases issued noncompetitively, the Mineral Leasing Act sets a 12.5 percent royalty rate by statute.

[32] GAO-08-691.

[33] According to Interior's *Fiscal Year 2014 Budget Justifications and Performance Information*, the administration proposed a package of legislative and administrative proposals to reform the management of Interior's onshore and offshore oil and gas leasing programs, with a key focus on improving the return to taxpayers from the sale of these federal resources and on improving transparency and oversight. Proposed changes included advancing royalty reforms such as evaluating minimum royalty rates for oil, gas, and similar products; adjusting onshore royalty rates; analyzing a price-based tiered royalty rate; and repealing legislatively-mandated royalty relief. The budget proposals for the past 4 years have included the goal of increasing onshore royalty rates.

[34] According to officials, Interior developed this estimate in support of the budget in 2011. The estimate is based on a royalty rate of 18.75 percent for oil and gas. Actual changes in

revenue collections resulting from a royalty rate increase would be highly dependent on market prices and production.

[35] GAO-08-691.

[36] Wood Mackenzie, *Government Take: Comparing the Attractiveness and Stability of Global Fiscal Systems* (June 2007).

[37] In the draft of our September 2008 report that we sent to Interior for comment, we made recommendations to address these issues. In its response, Interior stated that it did not fully concur with our recommendations because it had already contracted for a study that would address many of the issues we raised. However, because Interior's ongoing study was limited in scope and to a specific region in the Gulf of Mexico, rather than a review of the entire federal oil and gas fiscal system as we recommended, we did not find the department's stated rationale for not agreeing fully with our recommendations to be convincing. After Interior disagreed with our draft recommendations, for our final report we changed our recommendation to Interior into a "Matter for Congressional Consideration" that Congress should consider directing the Secretary of the Interior to (1) convene an independent panel to perform a comprehensive review of the federal system for collecting oil and gas revenue and (2) establish procedures for periodically collecting data and information and conducting analyses to determine how the federal government take and the attractiveness for oil and gas investors in each federal oil and gas region compare to those of other resource owners and report this information to Congress. Although Interior initially disagreed with this recommendation, in 2010, it contracted for the study comparing the federal oil and gas fiscal systems with those of other countries.

[38] GAO, *Mineral Revenues: MMS Could Do More to Improve the Accuracy of Key Data Used to Collect and Verify Oil and Gas Royalties*, GAO-09-549 (Washington, D.C.: July 15, 2009).

[39] 76 Fed. Reg. 30878 (May 27, 2011).

[40] GAO/AIMD-00-21.3.1.

[41] GAO-08-691.

[42] Office of Management and Budget, *Increased Emphasis on Program Evaluations*, M-10-01, Memorandum for the Heads of Executive Departments and Agencies (Washington, D.C.: The White House, Oct. 7, 2009).

[43] The Federal Onshore Oil and Gas Leasing Reform Act of 1987 requires that all public lands available for oil and gas leasing be offered first by competitive leasing. BLM is required to accept the highest bid received that exceeds the minimum bid value of $2 per acre or fraction thereof. 30 U.S.C. § 226(b)(1). The law allows the Secretary to increase the $2 an acre minimum bid and directs that the House and Senate Committees on Natural Resources be notified 90 days before doing so. By statute, the Secretary of the Interior must first offer parcels at competitive lease sales and may only issue noncompetitive leases after the department has offered the lands competitively at an oral auction and not received a bid. The annual rental rate is $1.50 per acre for the first 5 years and $2.00 per acre each year thereafter. Royalty rates for onshore are generally 12.5 percent for both competitive and noncompetitive leases. However, there are a few exceptions such as sliding scale royalties on older leases and reduced royalty rates on certain oil leases with declining production and reinstated leases. See 43 C.F.R. § 3103.2-3.

[44] Interior is prohibited by statute from evaluating the value of the lands proposed for lease onshore. 30 U.S.C. §226(b)(1). Concerned that BLM's onshore leasing system was not generating revenues comparable to what might be obtained through competitive leasing, Congress passed the Federal Onshore Oil and Gas Leasing Reform Act of 1987, which amended the Mineral Leasing Act of 1920. This act significantly changed the way BLM issues leases. Prior to the act, BLM was required to evaluate federal lands for oil and gas potential. The act requires the market, rather than administrative determinations, to set the value of leases by making all leases available for competitive leasing.

[45] GAO/AIMD-00-21.3.1.

In: Oil, Gas, and Coal Resources …
Editor: Marybeth T. Baldwin

ISBN: 978-1-63117-645-6

Chapter 2

COAL LEASING: BLM COULD ENHANCE APPRAISAL PROCESS, MORE EXPLICITLY CONSIDER COAL EXPORTS, AND PROVIDE MORE PUBLIC INFORMATION*

United States Government Accountability Office

WHY GAO DID THIS STUDY

In fiscal year 2012, about 42 percent of the 1.05 billion tons of coal produced in the United States came from coal tracts leased under the federal coal leasing program. Interior's BLM is responsible for managing this program, including estimating the fair market value of the coal to be leased. GAO was asked to examine this program. (Representative Markey originally made this request as Ranking Member of the House Committee on Natural Resources. He is now a member of the United States Senate.) This report examines (1) the number of tracts leased, along with the trends in associated coal production and revenues generated since 1990; (2) BLM's implementation of the process to estimate fair market value for coal leases; (3) the extent to which BLM considers coal exports and domestic coal reserve estimates when estimating fair market value; and (4) the extent to which BLM

* This is an edited, reformatted and augmented version of the United States Government Accountability Office publication, GAO-14-140, dated December 2013.

communicates information on federal coal lease sales to the public. GAO analyzed data on coal leasing activity, examined regulations and case files for coal lease sales, and interviewed BLM and other officials.

What GAO Recommends

GAO recommends, among other things, that BLM require state offices to use more than one approach to estimate fair market value where practicable, develop a mechanism to ensure that reviews of appraisal reports take place, and take steps to release additional summary information on its websites, including past lease sales. Interior concurred with these recommendations.

What GAO Found

Since January 1990, the Bureau of Land Management (BLM) has leased 107 coal tracts, and associated coal production and revenues have grown. Most lease sales have had a single bidder and were leased the first time offered. The amount of coal produced from federal leases and associated revenues have increased since 1990, although production has leveled off since 2002. Revenues from federal coal leases have generated about $1 billion annually in recent years. Royalties paid when coal is sold and bonus bids paid for the right to mine a federal coal tract account for nearly all of these revenues.

BLM's guidance offers flexibility in how to estimate fair market value, and BLM state offices vary in the approaches they used to develop an estimate of fair market value. In estimating fair market value, some BLM state offices used both the comparable sales approach—where bonus bids received for past sales are used to value the tract being appraised—and the income approach—which uses estimates of the future net revenue streams from the sale of coal from the appraised tract. However, some offices relied solely on the comparable sales approach and may not be fully considering future market conditions as a result. In addition, GAO found that BLM did not consistently document the rationale for accepting bids that were initially below the fair market value presale estimate. Furthermore, some state offices were not following guidance for review of appraisal reports, and no independent review of these reports was taking place. Adequate review of the fair market value process is critical to ensure that its results are sound and key decisions are fully documented. In addition, BLM is not currently taking advantage of a potential independent third-party reviewer with appraisal expertise within the Department of the Interior (Interior), specifically, the Office of Valuation Services.

BLM considers exports to a limited extent when estimating fair market value and generally does not explicitly consider estimates of the amount of coal that can be mined economically, known as domestic reserve estimates. As a result, BLM may not be factoring specific export information into appraisals or may not be fully considering the export potential of a lease tract's coal as called for in agency guidance. The Wyoming and Montana BLM state offices considered exports, but they generally included only generic statements about exports in the reports they prepared. In the other seven states with leasing activity, exports were generally not considered during the appraisal process. According to BLM officials, domestic reserve estimates, which vary based on market conditions and the costs to extract the coal, are not considered due to their variable nature.

BLM generally provides limited information on federal coal lease sales to the public because of the sensitive and proprietary nature of some of this information. The Wyoming BLM state office posts information on its website, including information on past lease sales, but most state office websites provide only general information. BLM's guidance states that redacted public versions of its appraisal reports should be prepared, but no BLM state office has prepared such reports. BLM supplied redacted versions of fair market value documents in response to a recent public information request only after being advised to do so by Interior's Solicitor's office.

ABBREVIATIONS

BLM	Bureau of Land Management
Btu	British thermal unit
EIA	Energy Information Administration
EPA	Environmental Protection Agency
FCLAA	Federal Coal Leasing Amendments Act of 1976
FOIA	Freedom of Information Act
IEA	International Energy Agency
NEPA	National Environmental Policy Act
OMB	Office of Management and Budget

ONRR Office of Natural Resources Revenue
USGS United States Geological Survey

December 18, 2013

The Honorable Peter DeFazio
Ranking Member
Committee on Natural Resources
House of Representatives

The Honorable Edward J. Markey
United States Senate

Coal is an important domestic energy source, and in 2011, coal-fueled electric power plants supplied about 42 percent of the nation's electricity. The Department of the Interior's Bureau of Land Management (BLM) is responsible for managing the coal resources on about 570 million acres of federal, private, and state land under the federal coal leasing program.[1] Under this program, BLM leases out federal coal tracts to mining companies who extract the coal from both surface and underground mines. In fiscal year 2012, about 42 percent of the 1.05 billion tons[2] of coal produced in the United States came from federal coal lease tracts.[3] The coal leasing program also generates significant revenue for federal and state governments; in fiscal year 2012, about $1.2 billion was generated from coal leasing.[4] These revenues come primarily from royalties paid on the coal when it is sold and payments made by companies to obtain the rights to mine on a federal lease tract, known as bonus bids.

Since 1990, all federal coal leasing has taken place through a lease-by-application process where coal companies propose tracts of land to be put up for sale by BLM.[5] At these sales, known as lease sales, companies can bid for the rights to lease tracts of land that contain federal coal for a set period of time; during the lease period, they can mine and sell coal from these tracts. In most cases, these lease tracts are adjacent to companies' existing coal mines, and the additional coal would allow these operations to continue. In preparation for a lease sale, BLM develops a confidential estimate of fair market value, which can generally be defined as the amount that a knowledgeable seller would obtain from a knowledgeable buyer for the coal deposit.[6] This estimate of fair market value is documented in an appraisal report prepared by BLM.[7] When conducting the lease sale, BLM leases the tract to the highest qualified bidder, as long as its bonus bid meets or exceeds BLM's estimate of fair market value.[8] In response to a lease application, BLM will also determine the amount of the coal that can be extracted from the lease tract and the environmental impacts of the proposed mining activity.

According to the Energy Information Administration (EIA),[9] coal exports have increased in recent years—particularly exports to Asia and Europe, where coal prices are generally higher than U.S. domestic prices. In 2012, the United States exported about 126 million tons of coal—an increase of 54 percent over 2010 levels. This recent increase in coal exports has raised questions about whether BLM's process for estimating fair market value is taking these changes into account, and whether the agency considers the total amount of coal that can be mined in the United States economically, known as domestic reserve estimates. In addition, some stakeholders, particularly environmental groups, have raised concerns about the amount of publicly available information on the federal coal leasing program and, specifically, documents BLM prepares as part of estimating fair market value. Interior's Inspector General also recently issued a report examining aspects of the federal coal leasing program, including the process for estimating fair market value and the coal lease inspection and enforcement program.[10]

You asked us to examine the federal coal leasing program. (This request was originally made by Representative Edward J. Markey as Ranking Member of the Committee on Natural Resources, House of Representatives. Mr. Markey is now a member of the United States Senate.) Our objectives for this report were to examine: (1) federal coal leasing, including the number of tracts leased, along with the trends in associated coal production and revenues generated since 1990; (2) BLM's implementation of the process to develop an estimate of fair market value for coal leases; (3) the extent to which BLM considers coal exports and domestic coal reserve estimates when developing an estimate of fair market value; and (4) the extent to which BLM communicates information on federal coal lease sales to the public.

To provide information on trends in federal coal leasing, we analyzed data from BLM's LR2000 database—which BLM uses to track federal land and mineral resources, including coal—and summarized federal coal lease sale activity and bonus bids accepted since 1990. We also analyzed data on coal production and revenues generated from federal coal leases from 1990 to 2012 from the Department of the Interior's Office of Natural Resources Revenue (ONRR), which is responsible for collecting and distributing revenues associated with federal mineral leases including federal coal leases. To assess the reliability of these data, we conducted interviews with BLM and ONRR officials regarding these data and reviewed documentation on their data systems, and we determined the data we used to be sufficiently reliable for our purposes, unless otherwise indicated.

To examine BLM's implementation of the process to develop an estimate of fair market value, we reviewed applicable regulations and BLM's guidance for the coal leasing program and interviewed BLM officials in both headquarters and state offices on how they implement these regulations and guidance. In addition, we reviewed appraisal standards developed by appraisal organizations in the United States and in other countries and spoke with officials from some of these groups. We also selected and reviewed a nonrandom sample of case files prepared by BLM officials as part of 31 coal lease sales using a data collection instrument we developed. Specifically, we focused on recent lease sales and examined case files for lease sales that generally took place from January 1, 2007, to July 31, 2012.[11] For those states that did not oversee a lease sale during this time frame, we examined files from their two most recent sales.

To determine the extent to which BLM considers coal exports when developing an estimate of fair market value, we used the results of our case file review to examine what types of information BLM included on exports. We interviewed BLM officials to learn about the information they consult in estimating fair market value and also spoke with knowledgeable stakeholders, such as academics, about future projections for coal exports. To determine the extent to which BLM considers domestic coal reserve estimates, we interviewed various BLM officials at headquarters and all of the BLM state offices where there are coal leases. In addition, we examined available export and domestic coal reserve information from government sources and coal mining companies.

To examine the extent to which BLM provides information to the public on coal lease sales, we analyzed BLM's policies for making information publicly available, reviewed BLM websites related to federal coal leasing, and reviewed a sample of documents that are made publicly available during the coal leasing process. We also interviewed BLM officials, representatives from industry, and environmental groups to get their perspectives on the information made publicly available on federal coal leases.

We conducted this performance audit from June 2012 to December 2013 in accordance with generally accepted government auditing standards. Those standards require that we plan and perform the audit to obtain sufficient, appropriate evidence to provide a reasonable basis for our findings and conclusions based on our audit objectives. We believe that the evidence obtained provides a reasonable basis for our findings and conclusions based on our audit objectives. A more detailed description of our objectives, scope and methodology is presented in appendix I.

BACKGROUND

Coal is an important domestic energy source, and BLM is responsible for managing coal resources on about 570 million acres of federal, state, and private land. Since 1990, all federal coal leasing has taken place through a lease-by-application process where companies propose lease tracts to be put up for sale by BLM. In fiscal year 2012, about 1.05 billion tons of coal was produced in the United States, including production from federal coal leases, and the biggest coal production area for federal coal was the Powder River Basin in northeast Wyoming and southeast Montana. Coal is also an important fuel source worldwide and consumption of coal continues to increase. To meet this growing demand, there has been an increase in global trade of coal, including exports from the United States.

Coal Leasing Program and the Transition from Regional Leasing to Lease-by-Application

The Federal Coal Leasing Amendments Act (FCLAA) of 1976 amended the Mineral Leasing Act of 1920 to generally require that all federal coal leases be offered competitively.[12] Competitive leasing provides an opportunity for any interested party to competitively bid for a federal coal lease. There are two procedures that can be used for competitive leasing: (1) regional leasing, where the Secretary of the Interior selects tracts within a region for competitive sale based on, among other things, expected demand for coal resources and potential economic impacts and (2) lease-by-application, where companies submit an application to nominate lease tracts that they are interested in leasing. Under both of these methods, BLM examines the potential environmental impact that could result from coal leasing.

In April 1982, the first regional coal lease sale was held for 13 lease tracts containing 1.6 billion tons of coal located in the Powder River Basin in Montana and Wyoming, and a follow-up sale was held in October 1982 for 2 lease tracts. Controversy surrounded the 1982 sale. Specifically, there were allegations that confidential appraisal information was disclosed to coal companies prior to the lease sale and that appraisal and sale procedures failed to assure that the public received fair market value for the leased coal tracts. These allegations led to an investigation by the House Appropriations Committee and a report that we issued in May 1983.[13] Later that year, Congress directed the Secretary of the Interior to establish a commission to review the coal leasing procedures to ensure the receipt of fair market value, known as the Commission on Fair Market Value Policy for Federal Coal Leasing or the Linowes Commission. Congress imposed a moratorium on lease sales until after the commission's final report was issued in 1984. Among its key findings on the fair market value process, the Linowes Commission found that Interior used appraisal methods that were widely accepted by industry and government, but that Interior needed to, among other things, enhance its capacity to perform appraisals and seek independent reviews of its appraisals and, more broadly, of the federal coal leasing program. From March 1984 through February 1987, coal leases were subject to another moratorium to enable development and implementation of revised coal leasing procedures based on the commission's recommendations.

By 1990, BLM had shifted from regional coal leasing to lease-by-application as the primary method of conducting federal coal lease sales. From 1987 to 1990, Interior decertified six coal regions it had established under the regional leasing program, citing declining interest in coal leases and poor coal market conditions.[14] Decertification meant that regional sales were no longer conducted, but that lease-by-application was available so that current mines could maintain production at their existing mines or new mines could begin operations. Under the lease-by- application process, companies may submit applications to BLM state offices to nominate lease tracts to be put up for sale.[15] This contrasts with the regional leasing process where Interior would decide which lease tracts would be put up for sale.

Tracts nominated under the lease-by-application process, commonly referred to as maintenance tracts, are generally adjacent to existing mining operations and are nominated by companies that own these operations. The BLM state office where the tract is located will review the application to determine whether it is consistent with applicable regulations, or if leasing the proposed property would be contrary to the public interest. For example, a lease application may be rejected if BLM determines that the land is unsuitable for coal mining or if a qualified surface owner does not consent to surface mining.[16] During this review process, BLM may also choose to redraw the lease tract boundaries in the public interest, a process known as tract modification. Reasons for tract modification include ensuring that economically recoverable coal adjacent to the original lease tract not be bypassed, or enticing another mining company to bid on a lease tract by making the boundaries of the proposed tract adjacent to more than one potential bidder, according to BLM officials. Once BLM accepts an application, it will begin either an environmental assessment or an environmental impact statement in accordance with the National Environmental Policy Act (NEPA).[17]

In preparing for a lease sale, BLM will also develop a presale estimate of fair market value of the lease tract's coal, which is generally expressed in cents per ton of coal that is recoverable from the lease tract. "Recoverable" refers to an estimate of the amount of coal that can be commercially mined from the tract and excludes coal that is not mined, such as top and bottom sections of a coal seam, which are typically mixed with less valuable rock.[18] There are also instances when fair market value is expressed on a per acre basis.

The presale estimate of fair market value is generally documented in an appraisal report prepared by the BLM state office overseeing the lease sale.[19] Other reports, such as geologic, engineering, and economic reports, may also be prepared during the appraisal process by either the relevant BLM state office or an associated BLM district or field office in the state.[20] The geologic report contains a legal description of the tract, along with an estimate of the amount of coal that can be recovered on the lease tract along with the characteristics of the coal, including its heating content.[21] An engineering report generally contains a mining plan, along with estimates of the costs to extract the coal based on the number of employees and capital equipment necessary to carry out this plan, among other costs. An economic report provides information on future coal market conditions, including price and demand levels for the lease tract's coal.

Prior to a lease sale, BLM is required to publicly announce in the Federal Register and a local newspaper when and where a lease sale will be held and the bidding procedures. Any company is free to bid on the lease using a sealed bid process. The amount that a company will pay to lease the tract—known as a bonus bid—is a function of the cents per ton they are willing to pay multiplied by the estimated recoverable tons of coal from a lease tract. These bonus bids are then reviewed by a BLM sales panel, which includes officials from the relevant BLM state office and BLM headquarters. Bids are accepted or rejected based on whether they meet the estimate of fair market value, and the lease is awarded to the highest qualified bidder that meets or exceeds this estimate of fair market value.[22] This successful bidder must either pay the total bonus bid in full at the time of lease sale or pay 20 percent of the bonus bid at the lease sale followed by four equal payments on the first four anniversary dates of the lease. The minimum bid that BLM can accept for a lease tract is $100/acre. If a lease sale does not receive a qualified bid at or above the estimate of fair market value, the lease tract can be renominated again through the lease-by-application process by the company that originally nominated the tract or by another interested company. If there is no interest in the lease tract, the application is closed by BLM.

In addition to paying a bonus bid for the rights to mine the coal on a lease tract, companies also pay rents and royalties on the coal they extract.[23] Rent amounts are at least $3 an acre and royalties are 8 percent of the sale price for coal produced from underground mines and at least 12.5 percent of the sale price for coal produced from surface mines. These royalties are paid on the price of the coal received at the first point of sale after it is removed from the ground. Tracts are leased out for an initial 20- year period, so long as the lessee produces coal in commercial quantities within a 10-year period and meets the condition of continued operations.

Lease terms can be extended if a company is actively producing coal on the lease tract.

U.S. Coal Production

According to EIA data, about 1.1 billion tons of coal was produced in the United States in 2011 from 1,325 mines, which employed over 91,000 people.[24] Coal is produced from three major regions—Appalachia, the interior United States, and the western United States (see Figure 1). More than half of U.S. coal came from the western region, which includes the Powder River Basin in northeast Wyoming and southeast Montana. The Powder River Basin is the largest coal-producing region in the United States, and all 10 of the top-producing U.S. coal mines are in the Powder River Basin, with 9 of these located in the Wyoming portion of the basin, according to EIA data. Coal in the Powder River Basin has less sulfur than eastern coals, making it attractive to utilities for meeting Clean Air Act requirements.[25] Close to 100 percent of federal coal is produced from leases located in the western region and, in fiscal year 2012, federal coal accounted for nearly 80 percent of the western region coal production totals. Production from the western region is expected to continue to be the largest source of coal production in the future—in 2040, an estimated 56 percent of total U.S. coal production will come from western mines according to our analysis of EIA data.

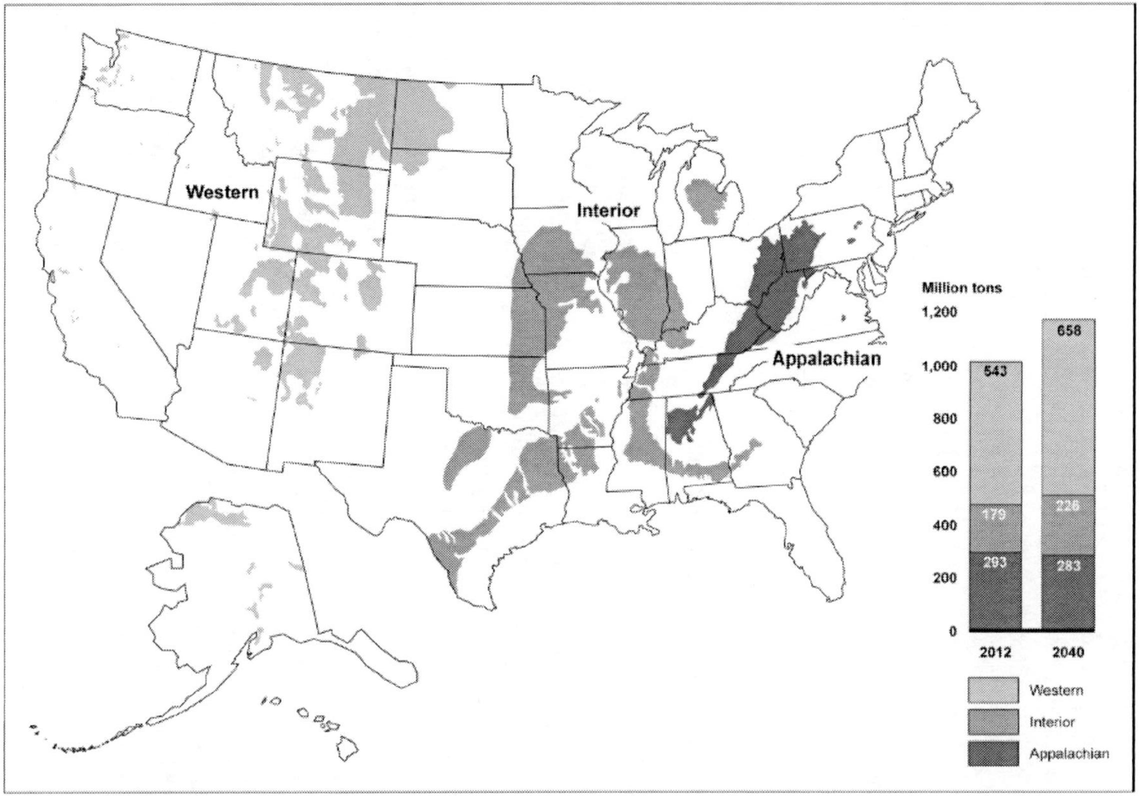

Source: GAO analysis of Energy Information Administration data;

Figure 1. U.S. Coal Production by Region.

There are two primary methods used to mine coal: underground mining and surface mining. Deeper coal resources require use of underground mining, which entails digging a series of mine entries and shafts and using equipment to extract the coal and transport it to the surface. Underground mining is more expensive than surface mining, which is used where coal deposits are buried within a few hundred feet of the surface. In surface mining, soil and rock above the coal—known as overburden—is blasted with explosives and removed using large equipment, and the uncovered coal is then extracted. Mining on federal leases involves both underground and surface mining. According to BLM officials, underground mining is generally used on federal leases in Colorado and Utah, and surface mining is generally used in Montana and Wyoming.

Market for Coal and Coal Prices

Domestically, coal continues to be an important energy source and fuels a large portion of the electric power sector in the United States, according to EIA data. In 2011, coal-fueled electric power plants supplied about 42 percent of the nation's total electricity and, within the past decade, coal has provided as much as 50 percent of electricity in the United States. More than 90 percent of the coal consumed in the United States is used by the electric power sector. According to EIA, for this reason, coal production trends are strongly influenced by coal demand in the electric power sector, which is sensitive to both changes in the overall demand for electricity generation and changes in the mix of fuel sources. Recently, there has been a general decline in the amount of coal used to generate electricity in the United States due to a combination of factors including a decline in overall electricity demand and shifts in the relative prices of other fuels.

Coal used in electricity generation is referred to as steam coal, as the coal is burned to produce steam which turns turbines that generate electricity. Most of the coal that is leased out through the federal leasing program is steam coal, according to BLM officials. In addition to its use in the generation of electricity, coal can also be used for a variety of industrial uses. For example, metallurgical coal is baked at high temperatures to make coke, which is used as fuel to make steel. Metallurgical coal has low sulfur and ash content, among other properties needed for making coke.

The amount of coal produced and consumed worldwide continues to increase. The International Energy Agency (IEA) reported that worldwide coal production increased by 6.6 percent in 2011, the twelfth straight year of growth.[26] In addition, as of 2011, coal supports 28 percent of the total primary energy consumption worldwide and is the second primary energy source behind oil. China continues to drive much of the world coal markets as its consumption and production of coal accounted for about 45 percent of both global consumption and production totals in 2011 according to IEA data. To respond to this growing international demand, there has been an increase in coal exports with global coal trade increasing 7 percent in 2011 according to IEA.

The United States exports a small but increasing amount of coal primarily to Europe and Asia and, in 2011, the United States ranked fourth globally in coal exports behind Indonesia, Australia, and Russia. According to EIA data, total U.S. coal exports more than tripled from 2002 to 2012, as shown in figure 2 below. In 2012 about 126 million tons of coal was exported—about 12 percent of the total coal produced in the United States. The majority of this coal is exported to Europe and Asia. Metallurgical coal, which is generally not mined on federal coal leases, has historically made up the majority of U.S. coal exports. Nonetheless, there has been growth in exports the last few years of steam coal—the primary type of coal mined on federal coal leases. Specifically, from 2010 to 2012, steam coal exports from the United States more than doubled, rising from 25.6 million tons to 55.9 million tons. Based on EIA data, exports from Wyoming and Montana, the two largest states in terms of production from federal leases, accounted for less than 2 percent of total U.S. coal production in 2011. In addition, coal companies have announced plans to further increase steam coal exports in the future, and there are several coal export facilities that are being proposed on the West Coast to transport coal to growing Asian markets.

The price for coal varies widely across the United States. Among the four states with the most production from federal coal leases—Colorado, Montana, Utah, and Wyoming—the average prices for coal originating in these states in 2011 were $39.88/ton in Colorado, $16.02/ton in Montana, $33.80/ton in Utah, and $13.56/ton in Wyoming, according to EIA's 2011 *Annual Coal Report.*[27] This large difference in price is tied to coal quality, which is referred to as coal rank.[28] Among other factors, coal rank is determined by the amount of carbon that the coal contains and the amount of heat energy it can produce, with higher rank coal having more energy content. The total amount of coal that an electric utility will need to fuel a power plant is tied to the heat content of coal. For example, a utility will need to buy more tons of coal with lower energy content to achieve the same output of energy that could be attained using less coal with a higher energy content. Other factors that affect a coal's quality are sulfur, moisture, and ash content. The sulfur content of the coal affects the sulfur dioxide emissions that result when coal is burned, and using coal with less sulfur content can help electric utilities meet air quality requirements. Coal with higher moisture and ash content is lower rank because both of these impact the amount of energy obtained from burning the coal. For example, coal with lower moisture content has greater energy content.

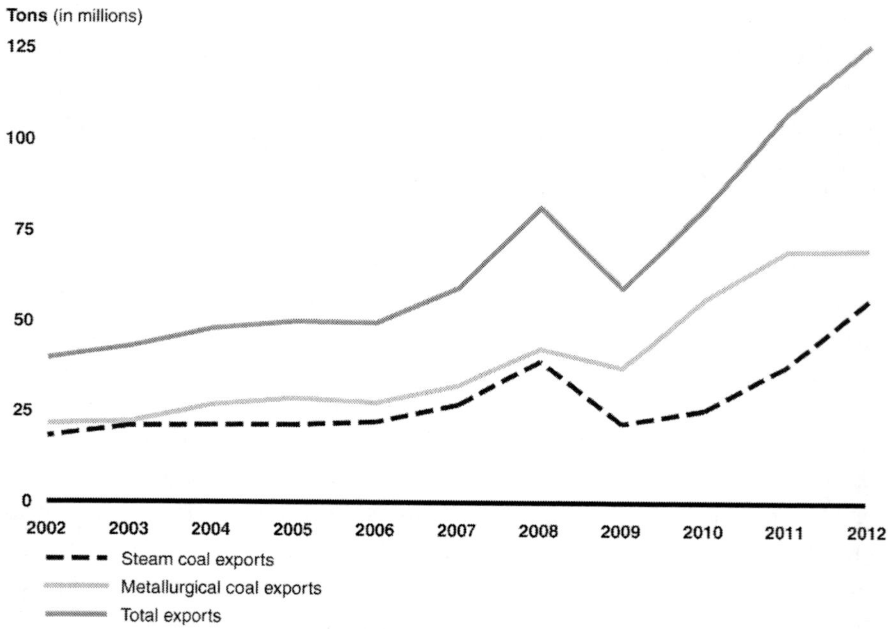

Source: GAO analysis of Energy Information Administration data.

Figure 2. U.S. Coal Exports, 2002-2012.

SINCE 1990, OVER 100 COAL TRACTS HAVE BEEN LEASED, AND COAL PRODUCTION AND ASSOCIATED REVENUES HAVE GENERALLY GROWN

Since January 1990, BLM has leased 107 coal tracts under the lease-by-application process, and both coal production and the associated revenues have grown. Most lease sales had a single bidder, and the successful bid amounts—typically expressed in cents per ton—have varied by state, with the greatest increases over time observed in Wyoming. The amount of coal produced from federal leases and associated revenues increased from fiscal year 1990 to fiscal year 2002.[29] Since fiscal year 2002, coal production from federal leases has remained relatively steady, but revenues continued to grow. In total, revenues from federal coal leases have generated about $1 billion annually in recent years.

Table 1. Number of Federal Tracts Leased and Associated Amount of Coal by State from 1990 to 2012

State	Number of federal tracts leased	Acres	Tons of recoverable coal (Millions)	Percentage by tons of recoverable coal
Wyoming	31	77,137	7,967[a]	88.4%
Colorado	20	26,375	289	3.2%
Utah	15	30,082	214	2.4%
Montana	4	5,349	187	2.1%
North Dakota	12	8,386	135	1.5%
New Mexico	3	10,926	93	1.0%
Oklahoma	9	16,339	71	0.8%
Alabama	6	11,097	44	0.5%
Kentucky	7	2,952	10	0.1%
Total	**107**	**188,243**	**9,011**	**100.0%**

Source: GAO analysis of BLM data.

[a] In Wyoming, we are reporting primarily mineable tons of coal while, for the other states, we are reporting recoverable tons of coal. This estimate of mineable tons of coal is generally a larger number than the recoverable estimate, because it includes coal that is generally left in place during actual mining operations, such as coal along property boundaries or coal left in place as pillars for structural reasons in an underground mine. Wyoming BLM typically does not report recoverable tons publicly because officials in this state office consider this sensitive information.

Of the 107 Tracts Leased Since 1990, About 90 Percent Had a Single Bidder, and Most Were Leased the First Time Offered

In 1990, BLM began using the lease-by-application process as the primary method to lease out coal, and since then BLM has leased 107 coal tracts, 31 of which were in Wyoming. (See app. II for a complete list of lease sales held since 1990.) The coal from the Wyoming lease tracts comprise approximately 8 of the 9 billion tons, or about 88 percent, of the coal available from federal tracts leased since 1990, as shown in table 1.

Of the 107 leased tracts, sales for 96 (about 90 percent) involved a single bidder (see Figure 3), which was generally the company that submitted the lease application. More than 90 percent of the lease applications BLM received were for maintenance tracts used to extend the life of an existing mine or to expand that mine's annual production.

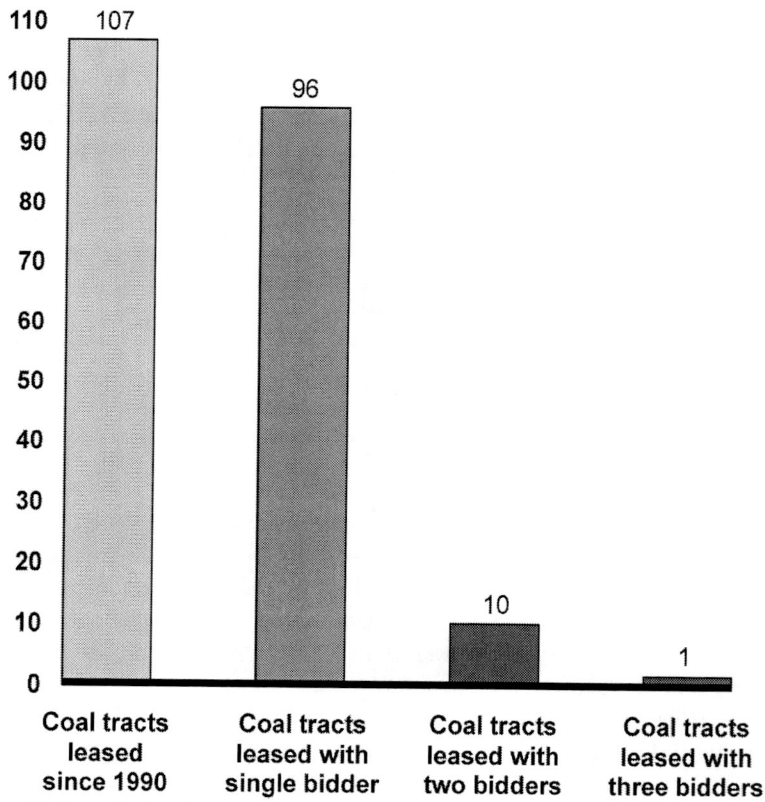

Source: GAO analysis of BLM data.

Figure 3. Number of Bids Received for Federal Coal Tracts Leased, 1990-2012.

According to BLM officials and coal industry representatives, there is limited competition for coal leases because of the significant capital investment and time required to establish new supporting infrastructure to start a new mine or to extend operations of an existing mine to a tract that is not directly adjacent to it. For these reasons, there have not been many new mines established on federal leases recently. For example, according to BLM officials the last new mine started on a federal lease in the Powder River Basin in Wyoming was the North Rochelle mine, which began operations in 1982. Officials from coal companies told us they typically submit new applications for federal coal leases to maintain a 10- year coal supply at their existing mining operations. In 1983, we noted a similar lack of competition for federal coal leases following the 1982 regional coal lease sale in the Powder River Basin and concluded that the market for coal leasing was largely noncompetitive because lease tracts sold "appear captive to adjacent mining operations."[30]

According to BLM officials, this same issue remains relevant today, and it is difficult to attract multiple bidders on a lease tract if it is not adjacent to multiple mining operations. For example, as shown in figure 4, tracts submitted for lease-by-application that are north and west of the Black Thunder mine are less likely to be bid on by the operators of the North Antelope Rochelle or Antelope mines. This is because it would be too costly and take significant time for these mine operators to move their heavy equipment to extract coal from these lease tracts, which are not directly adjacent to their existing operations. In contrast, the lease tracts that are located between two mines are more likely to be bid on by multiple mine operators, according to BLM officials.

BLM officials told us that, where possible, BLM uses the tract modification process to encourage competition for lease sales. For example, Wyoming BLM officials told us that they recently divided an applicant's proposed tract into two distinct tracts to be sold in two separate coal lease sales upon realizing that one segment may potentially interest another mining company.

Colorado BLM officials told us that they altered boundaries of one coal lease application to allow for multiple entry points to the coal for underground mining to make the tract attractive to other companies. In our review of case files related to 31 recent lease sales, we found that BLM modified boundaries for seven tracts (23 percent) to enhance competition.[31] Six of these tracts were located in Wyoming and comprised more than half of the 11 Wyoming lease sales we reviewed; 1 was located in Utah. None of these leases, however, received multiple bids when sold.

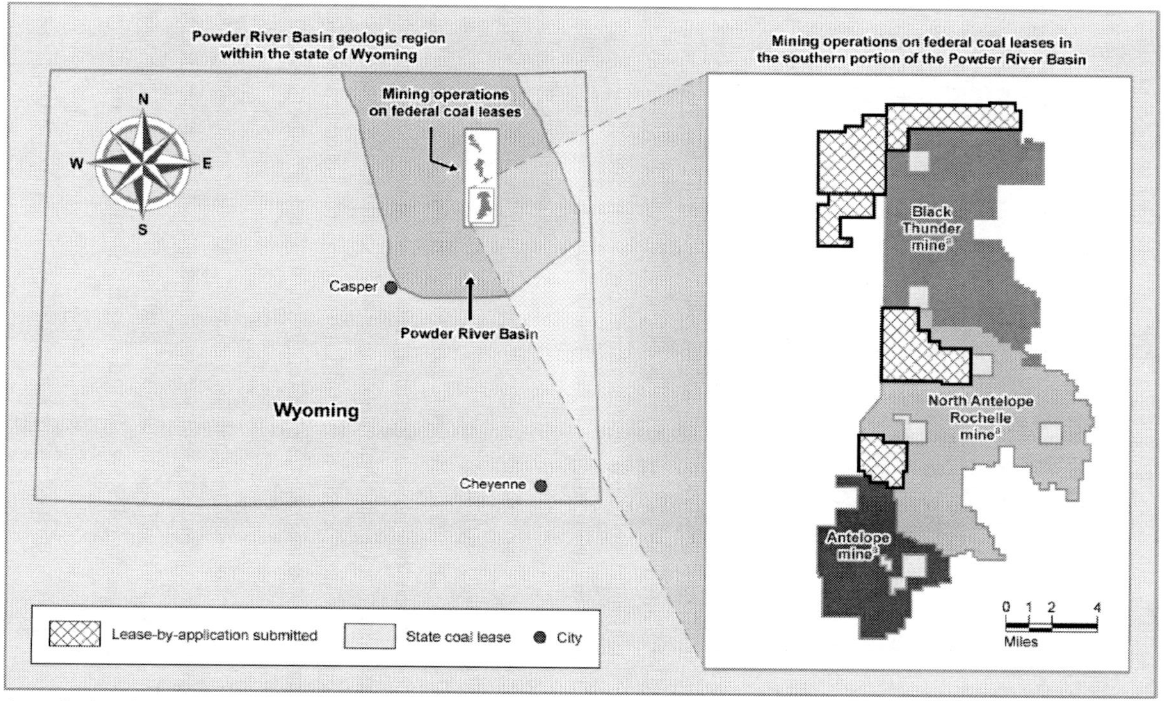

Source: GAO analysis of BLM information.

[a] This map is a general representation of coal mining operations in the Powder River Basin. The Black Thunder, North Antelope Rochelle, and Antelope mines are each composed of several lease tracts, but we have combined this into a single entity for representation purposes.

Figure 4. Map of Powder River Basin Coal Operations on Federal Coal Leases.

Of the 107 leased tracts, 89 (about 83 percent) were leased the first time they were offered for sale. According to representatives of appraisal organizations we spoke with, this high acceptance rate of initial bids may reflect the reliance of existing mines on federal coal leases to maintain their operations and a willingness of mine owners to submit slightly higher bids to ensure they win federal coal leases. The remaining 18 tracts were leased after being reoffered for sale one or more times because the initial bonus bid offered was below the estimate of fair market value.[32] Of the 18 tracts that were reoffered for sale, 8 were in Wyoming and 5 were in Colorado.

Amount of Coal Produced from Federal Leases Gradually Increased in the 1990s Before Leveling Off in Fiscal Year 2002, with Most Production Taking Place in Wyoming

The total amount of coal produced from federal leases has nearly doubled since fiscal year 1990. Growth in coal production from federal coal leases was largest from fiscal years 1992 to 2002, when it grew from 239 million tons to 444 million tons. The proportion of coal produced from federal leases relative to the total amount of U.S. coal production also grew over this same period from about 24 percent in fiscal year 1992 to about 40 percent in fiscal year 2002 (see Figure 5). During this period there was an increase in U.S. western coal production, where a majority of federal coal is located, and a corresponding decline in production from eastern coal regions. In particular, BLM officials told us that Powder River Basin coal grew in demand over eastern coal because it enabled utilities to meet the stricter emissions limits due to its low sulfur content. Powder River Basin coal was also attractive to utilities because of its low production costs and access to transportation networks, both of which help to decrease the market price that a utility must pay for the coal. A United States Geological Survey (USGS) study reported that this shift reflected the fact that western mines, which typically rely on surface mining, can extract coal more cheaply than eastern mines, where coal is generally mined using underground methods.[33]

Since fiscal year 2002, coal production from federal leases remained relatively steady, averaging near 450 million tons annually, or about 41 percent of total U.S. production. Production peaked in fiscal year 2008 at 483 million tons and has since declined by 8 percent to 442 million tons in fiscal year 2012. In October 2012, we reported the amount of electricity generated using coal has decreased recently due to a decline in overall electricity demand and growth in the use of natural gas to fuel power plants.[34]

In fiscal year 2012, 85 percent of the coal produced from federal leases came from Wyoming. As shown in figure 6, Wyoming and three other western states—Montana, Colorado, and Utah—accounted for 97 percent of coal produced from federal leases. The remaining 3 percent of coal (about 12 million tons) was produced from federal leases in five other states—Alabama, Kentucky, New Mexico, North Dakota, and Oklahoma.

Revenue from Federal Coal Leases Has Nearly Doubled Since 2003 and Generated About $1.2 Billion in Fiscal Year 2012

The total revenue generated from federal coal leases has nearly doubled from $682 million in fiscal year 2003 to $1.2 billion generated in fiscal year 2008 and again in fiscal year 2012.[35,36] Total revenues from federal coal leases have remained relatively steady since fiscal year 2005 averaging about $1.0 billion per year according to our analysis of ONRR data. There are three sources of revenue from federal coal leases— royalties, bonus bids, and rents––but royalties and bonus bids account for nearly 100 percent of the revenues from the federal coal leasing program.

Royalties. Royalties comprised the majority of the revenue from federal coal leases—nearly two-thirds of the total revenue over the period from fiscal years 2003 to 2012.[37] Royalty rates for coal depend on the mine type and are generally calculated based on a proportion of sales value, less allowable deductions, such as transportation and processing allowances.[38] BLM generally sets royalty rates at 12.5 percent for surface mines, the required minimum royalty rate, and 8 percent for underground mines, the rate prescribed by regulation.[39] In total, royalties generated from federal coal leases have more than doubled since fiscal year 1990, from $392 million to $796 million in fiscal year 2012 (see Figure 7). In addition, as with coal production from federal leases, royalties generated from the sale of coal from federal leases in Wyoming comprise an increasing proportion of the royalty stream ranging from 50 percent of total royalties in 1990 to 80 percent in 2012 (see Figure 6).

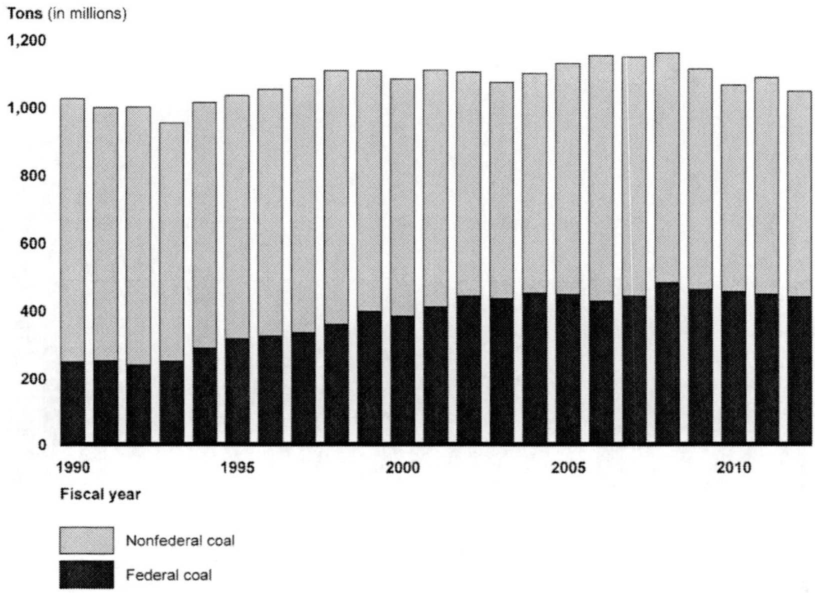

Sources: GAO Analysis of Office of Natural Resources Revenue and Mine Safety and Health Administration data.

Note: Non-federal coal sources include private, state, and tribal coal.

Figure 5. Share of U.S. Coal Produced from Federal Leases, Fiscal Years 1990-2012.

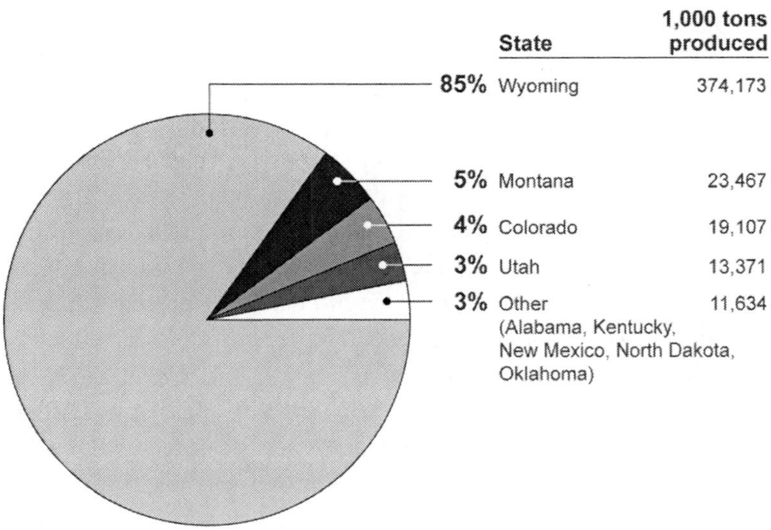

Source: GAO analysis of Office of Natural Resources Revenue data.

Figure 6. Coal Produced from Federal Leases by State, Fiscal Year 2012.

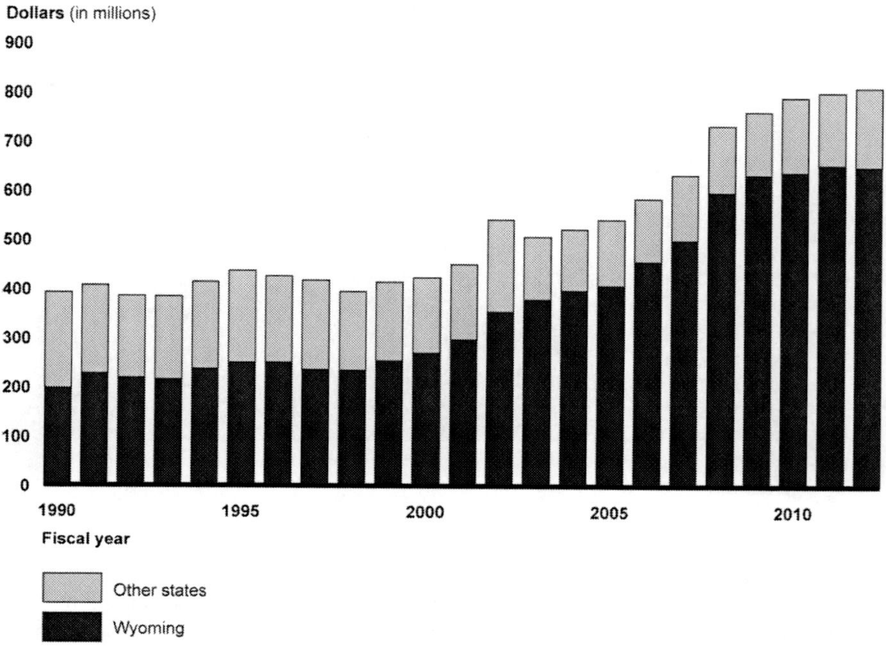

Source: GAO analysis of Office of Natural Resources Revenue and Energy Information Administration data.

Figure 7. Total Royalties Generated from Federal Coal Leases, Fiscal Years 1990 to 2012 (2013 dollars).

Coal prices have been a major driver of the increases in royalty revenues. For instance, from fiscal years 1990 to 2000, royalty revenues remained relatively steady even though production of federal coal increased over this period related to a decline in coal prices. Since then, coal royalty revenues have steadily increased, even with a recent decline in production. Specifically, from fiscal years 2008 to 2012 the amount of coal produced from federal leases declined by about 41 million tons of coal (or 8 percent); however the reported sales value of this coal increased 15 percent from $6.7 billion to $7.7 billion, reflecting growth in coal prices.

The effective royalty rate—the rate actually paid by lessees after processing and transportation allowances have been factored in along with any royalty rate reductions—generated from coal produced from federal leases has remained on average at about 11 percent since fiscal year 1990. Royalty rate reductions may be approved by BLM in cases where a reduction is needed to promote mining development. For example, BLM officials told us they may approve royalty rate reductions to enable continued operations in cases where mining conditions may be particularly challenging and costly, or to enable expanded recovery of federal coal. The effective royalty rate varies by state due to differences in mine type and other factors. For example, the effective royalty rate is higher in Wyoming and Montana where most coal is extracted using surface mining. In fiscal year 2012, the effective royalty rates for the top federal coal producing states were: Wyoming (12.2 percent), Montana (11.6 percent), Utah (6.9 percent), and Colorado (5.6 percent).

Bonus bids. Bonus bids are generally expressed in cents per ton of coal that is recoverable from the lease tract. The total bonus bid paid is the cents per ton multiplied by the estimated recoverable tons of coal from the lease tract. According to BLM officials, typically an initial payment of 20 percent of the total bonus bid is provided with the sealed bid, and the remaining 80 percent is paid in four equal annual installments over a 4- year period, but it may also be paid in full by the lessee at the time of a lease sale. ONRR revenue data from fiscal years 2003 to 2012 show total bonus bids received from all federal coal leases averaged $335 million annually, or about one-third of the total revenues from federal coal leases, as shown in figure 8. Since fiscal year 2003, revenue from bonus bids has fluctuated from year to year related to lease sale activity. For example, since fiscal year 2003, revenue from bonus bids has fluctuated from a peak of about $521 million in fiscal year 2005, when bonus bids made up 49 percent of the total revenue generated from coal leases, to a low of $116 million in fiscal year 2010, when bonuses comprised 13 percent of total revenue.

Based on our analysis of BLM data on coal lease sales, BLM accepted $6.4 billion in total bonus revenue for the 107 tracts leased since 1990, with total bids ranging from $5,000 to more than $800 million for a lease tract. In addition, successful bonus bid amounts for coal leases varied across states, with bonus bids received in Wyoming showing the greatest increase since 1990 when compared with the other seven states with active federal coal leases. Successful bonus bids for lease sale tracts in Wyoming ranged from $0.04 to $1.37 per ton of coal, after adjusting for inflation, and generally increased from 1990 to 2012. In comparison, successful bonus bids in Colorado bids ranged from $0.02 to $0.55 per ton and slightly increased from 1990 to 2012, and in North Dakota all successful bonus bids were $100 per acre in nominal dollars, the minimum bid BLM can accept for a lease tract and did not vary meaningfully over time when measured on a per ton scale. In other states, trends in bonus bids were not discernable due to variation in the successful bids over time or there being too few sales in these states. According to officials from coal companies we spoke with, bonus bids for federal coal leases depend on many factors, including coal quality, mine type (e.g., underground or surface mining), and the price of coal at the time of the sale.

Even when coal quality, mine type, and price are similar, successful bonus bids can vary greatly because of other factors. For example, mining conditions in Colorado and Utah are similar in several respects— most mines are underground, the energy content of the coal being mined generally exceeds 11,500 BTUs per pound of coal, and coal prices were in a similar range from 1990 to 2011.[40] Yet, the total bonus bids accepted by Colorado since 1990 have been about $22 million less after adjusting for inflation than those accepted by Utah despite the fact that Colorado has leased out almost 76 million tons more coal than Utah. When asked about the differences in total bonus bids, BLM officials reiterated that differences in conditions affecting coal marketability across these states, such as access to transportation options and proximity of customer base, make direct comparison of bonus bid values across these states difficult. Specifically, BLM officials told us that most of the coal produced in Utah is consumed locally by power plants in state; this proximity to the customer could be considered an advantage. In contrast, much of the coal produced in Colorado needs to be transported out of state.

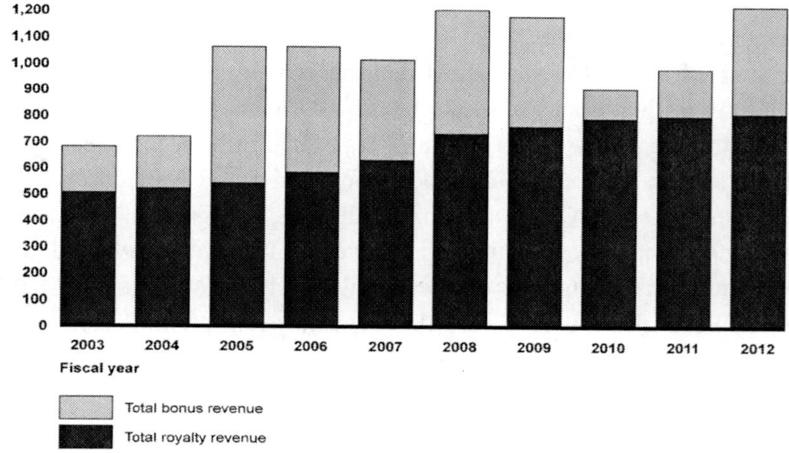

Source: GAO analysis of Office of Natural Resources Revenue data.

Note: Rents and other income account for less than 1 percent of the revenues collected from federal coal leases and are not shown in the above figure.

Figure 8. Bonus and Royalty Revenue Generated from Federal Coal Leases, Fiscal Years 2003-2012 (2013 dollars).

Rents. Rents, which are set at $3 per acre, are also collected annually from federal coal leasing tracts but comprise an insignificant amount of the revenue stream.[41] In fiscal year 2012, $1.4 million in rent revenue was generated from federal coal leases, composing 0.1 percent of the annual revenue related to coal.

BLM'S IMPLEMENTATION OF THE FAIR MARKET VALUE PROCESS LACKS SUFFICIENT RIGOR AND OVERSIGHT

BLM's guidance offers flexibility in how to estimate fair market value, and BLM state offices vary in the approaches they use to develop an estimate of fair market value. Some state offices use both the comparable sales and income approaches in their appraisals while others rely solely on the comparable sales approach and may not be fully considering future market conditions as a result. In addition, we found that BLM did not consistently document the rationale for accepting bids that were initially below the fair market value presale estimate, and some state offices were not following guidance for review of appraisal reports. Furthermore, no independent review of appraisals is taking place, as is recommended by commonly used appraisal standards, despite Interior having expertise that could be leveraged to do so.

BLM's Guidance Offers Flexibility in How to Estimate Fair Market Value

According to BLM guidance, the goal of BLM's appraisal process is "to provide a well-supported estimate of property value that reflects all factors that influence the value of the appraised property," and it gives state offices flexibility in how they do so.[42] BLM's guidance lays out two approaches to develop an estimate of fair market value—comparable sales and income—but does not say that both approaches must be used.[43] Under the comparable sales approach, bonus bids received for past sales are used to value the tract being appraised. Adjustments may be made to these comparable sales based on how the characteristics of these past lease tract sales compare with the lease tract being appraised. For example, if a past lease sale involved coal that had lower heating content than the lease tract being appraised, BLM might conclude that the current tract should have a higher fair market value than the bonus bid received for this past sale. In contrast, under the income approach, the revenues received from selling the coal and costs to extract it are projected into the future, and this net revenue stream is discounted back to the present. The resulting net present value of this revenue stream becomes an estimate of the fair market value for the lease tract. See table 2 for a summary of methods used and information needed for the comparable sales and income approaches.

BLM's guidance states that the comparable sales approach is preferred to the income approach when similar comparable sales are available because it is assumed that this method will provide the best indication of value. When comparable sales are not available, the guidance states that the income approach is a viable alternative, but the guidance highlights the uncertainty associated with using the income approach. This uncertainty stems from its reliance on projections of future market conditions, such as demand for coal, coal prices, and the costs to extract the coal. The guidance also provides examples for how the results of the comparable sales and income approaches can be used together. For example, information from comparable sales can be used as a comparison point for results from the income approach. In addition, results from the income approach can be used to adjust past comparable sales. Specifically, if the net present value of the tract being appraised is less than the net present value of a past lease sale, a conclusion can be made that the tract being appraised is less valuable than the past lease, and a numeric adjustment can be made to the actual sales prices of the past lease sale to account for this difference.[44]

Table 2. Summary of Approaches BLM Uses to Estimate Fair Market Value for Federal Coal Lease Tracts

Appraisal approach	General methods and steps	Information needed
Comparable sales approach	Uses past sales prices of coal tracts to estimate fair market value for tract being appraised. Compares characteristics of past lease tracts to tract being appraised to identify the most applicable sale(s) for use in the analysis and determines if any adjustments should be made to past sales prices. Reconciles, as necessary, results of the most applicable sale(s) and uses these results to estimate the fair market value.	Bonus bids paid in prior coal lease sales. Characteristics of lease tracts sold in prior transactions, such as: • Time of sale. • Coal quality (heating content value, sulfur content, ash content). • Type of mining to be used (surface or underground). • Physical characteristics of mining (i.e., depth to deposit, seam thickness). • Market conditions at time of sale.

Appraisal approach	General methods and steps	Information needed
Income approach	Estimates future net revenues from the sale of coal extracted from the lease tract using annual costs and revenue projections over the period of time that a deposit is expected to be mined, which could be more or less than 20 years, corresponding to the length of time that leases are initially issued for. Discounts, or converts, the future net revenue streams back to a single number to the present—referred to as the net present value. (BLM's guidance suggests using a 10 percent real discount rate to determine this net present value.) Uses the calculated net present value of the projected after-tax net revenue of the mine operation to estimate the fair market value of the lease tract. Determine whether and how to incorporate uncertainty surrounding future market conditions into the analysis.	Mine plan based on geologic and engineering data. Lease development plan and coal production schedule based on the mine plan. Coal price projections. Anticipated capital and operating costs, taxes, and other expenses for extracting the lease tract's coal.

Source: GAO summary of BLM guidance.

BLM State Offices Differ in Their Appraisal Approaches, and Some Offices May Not Be Fully Considering Future Market Conditions as a Result

During our interviews with BLM officials, we found that BLM state offices use different approaches to develop an estimate of fair market value of coal leases, and we confirmed this during our case file review. For example, for lease sales in Wyoming, Montana, and New Mexico,[45] the BLM state offices use both the comparable sales and income approaches, based on our review of case files. Moreover, the BLM Wyoming state office goes a step further to numerically adjust its comparable sales using the results of the income approach. In contrast, for lease sales in Colorado, North Dakota, Oklahoma, and Utah, the BLM state offices have generally used just the comparable sales approach in recent years. For the two lease sales we reviewed in both Alabama and Kentucky, one of the sales used both approaches, while the other used just the comparable sales approach. When using the comparable sales approach, BLM state offices generally only used sales information for coal sales that occurred in their state. (See app. III for specific information on the approaches used for the lease sales that we reviewed.)

BLM officials in some state offices said that they did not have the resources to perform appraisals using the income approach. In particular, the income approach may require the help of an economist, and some BLM state offices do not have an economist on staff. For example, officials in both the Utah and Colorado state offices said they did not have economists on staff. For this reason, the Utah BLM office recently contracted with a firm to help them perform the income approach for a lease under consideration. However, BLM headquarters officials told us that the income approach did not require an economist and that some mining engineers in state offices could perform appraisals using this method. Officials in other state offices said they could not justify using the income approach due to the market for coal in their states. For example, they said that most coal mining in Oklahoma involves privately held coal, and a bonus bid is not required to obtain the rights to mine the coal, while in North Dakota, bonus bids offered as part of private sales have generally been less than or equal to the $100/acre minimum required for federal coal leases. When using these private sales as comparable sales, BLM officials in these states concluded that the minimum bonus bid of $100/acre should be the estimate of fair market value. BLM officials told us that if they did not set fair market value at this level, the coal on the federal lease tracts would be bypassed and never mined.

The reliance solely on the comparable sales approach among certain BLM state offices contrasts with the recommendations of officials from appraisal organizations we spoke with, who generally supported using both the comparable sales and income approaches when conducting mineral valuations. Representatives from three U.S. appraisal organizations told us that the income approach can provide helpful information and should be used along with the comparable sales approach.[46] Specifically, the income approach can serve as a check on the results of the comparable sales approach. In addition, we reviewed general appraisal standards in the United States and industry-developed standards for mineral valuation in Canada and Australia, as identified by appraisal organizations we spoke with, and we found that mineral valuation standards in Canada were the most prescriptive in terms of using multiple appraisal methods.[47] Specifically, the Canadian standards require that more than one appraisal approach be used unless justification is provided, and these standards recommend use of both the income and comparable sales approaches.[48] All of the standards we reviewed stated that appraisal reports should include a discussion of the rationale for the appraisal approaches used, as well as the rationale for any approaches not used. Similarly, representatives from one of the appraisal groups we interviewed said that if only a single approach is employed, the reasons for doing so should be documented and justified. According to BLM's guidance, officials must document the rationale for choosing a certain appraisal approach in the appraisal report but, during our review of case files, we generally did not find this rationale documented in states where one approach was used. In contrast, appraisal reports prepared for lease sales in New Mexico, North Dakota, Montana, and Wyoming contained explanations for the appraisal approaches they chose to use.

Because the income approach examines estimates of future market conditions while the comparable sales approach focuses on past coal lease sales, BLM state offices that rely solely on the comparable sales approach may not be fully considering current or new trends in coal markets when estimating fair market value. This is particularly true if a state office is using comparable sales from a time during which market conditions were different. During our case file review, we found there were several comparable sales used that were over 5 years old. One official from an appraisal organization told us that he would hesitate to use comparable sales that were older than 5 years because of changes in market conditions. BLM officials noted that the usefulness of sales over 5 years old would depend on the extent to which the market has changed.

BLM Did Not Always Document the Rationale for Accepting Bids and Inconsistently Prepared Appraisal Reports

During our case file review of 31 selected lease sales, we found four lease tracts in three states where the bonus bid offered was below the fair market value presale estimate, but BLM accepted these bids after additional consideration was given to them.[49] In total, the accepted bonus bid amounts related to all four tracts was more than $2 million below the presale estimate of fair market value. Three of these sales occurred in the 1990s, and one occurred in 2007. As outlined in BLM's guidance, bonus bids below the presale estimate of fair market value may be considered as long as the bid is above the minimum bonus bid requirement of $100 per acre, among other factors. Furthermore, BLM's guidance allows for additional information to be considered or additional analysis to be completed as part of a postsale review process to address technical errors or in cases where appraisal standards are not met. BLM's guidance states that postsale analysis be documented and any revised fair market value be reviewed, but it does not clearly describe what postsale documentation is needed. According to BLM headquarters officials, this postsale analysis must be documented and a new estimate of fair market value needs to be completed and reviewed. We did not, however, find this documentation in the case files we reviewed for these four sales. Specifically, we found no documented evidence of a single, revised fair market value estimate against which to compare the bids. The files contained general statements about additional information that was considered during the postsale review process, such as changes in mining plans or changes in coal prices. In each of the four cases, BLM found that the respective bids fell within an "acceptable range of values" close to the initial presale fair market value estimate and, as a result, BLM determined in each of these cases that the bid should be accepted. Without better documentation of these decisions, including specifying the revised fair market value estimate and clear justification for the revision, BLM has not demonstrated that the accepted bids met or exceeded the fair market value estimate as required under the Mineral Leasing Act.

We also found inconsistencies in the appraisal reports prepared as part of coal lease sales. In particular, some states consistently updated past comparable sales for inflation while others did not. For example, we found instances where the Montana/Dakotas and New Mexico state BLM offices used comparable sales that were more than 5 years old, but did not adjust them for inflation. In contrast, the Colorado, Utah, and Wyoming BLM state offices generally updated sales that were more than 5 years old for inflation. BLM headquarters officials told us that past comparable sales should be adjusted for differences in market conditions over time. State offices also varied in the number of comparable sales they consulted when using the comparable sales approach. For the 31 lease sales we reviewed, the number of comparables used in the appraisal ranged from a low of 2 to a high of 10 comparable sales.

In addition, we found instances where BLM did not fully document its estimate of fair market value. Specifically, we found three related lease sales in Oklahoma where a formal appraisal report was not prepared to justify using the minimum bid amount of $100/acre as the estimate of fair market value. In the case file, there was discussion of the general market for coal in Oklahoma, including the fact that private coal sales did not involve up-front payments, such as bonus bids, but there was no description of the methods used to develop an estimate of fair market value. A BLM official said that he believed comparable sales were reviewed to determine that the fair market value estimate would be below the minimum bid value for these leases, but this was not documented in a formal appraisal report.

Some State Offices Did Not Follow BLM's Guidance for Reviewing Appraisal Reports and Currently There Is No Independent Review of Appraisals

From our review of 31 case files, we found differences in the appraisal review process used by different state offices and, in some cases, states had not followed BLM guidance. According to BLM guidance, appraisal reports must be signed by three BLM officials—the chief of the regional evaluation team, a qualified mineral reviewer, and the deputy state director—to ensure technical accuracy of the fair market value estimate and conformance with BLM's appraisal guidance. The chief of the regional evaluation team is an outdated position that no longer exists because BLM no longer leases coal on a regional basis, but the guidance has not been updated to reflect this. BLM headquarters officials said they expected that the mineral appraiser's signature would take this official's place. However, we found that appraisal reports were not consistently signed by the three officials, and there was no mechanism in place to ensure that this review was taking place. While appraisal reports in Wyoming were signed by three officials—the mineral appraiser, mineral reviewer, and deputy state director—other state offices had appraisal reports that were reviewed and signed by a single official.[50] For example, two appraisal reports in Colorado were signed only by the branch chief of solid minerals, while in Alabama, one appraisal report was just signed by an economist. Of the two appraisals we reviewed for lease sales in Kentucky, one was signed by only an economist, and one was not signed at all. Without clear guidance on who is supposed to be reviewing reports and consistent reviews by these officials, BLM does not have assurance that proper oversight is taking place in all state offices responsible for coal leasing.

Currently, review of appraisal reports takes place primarily at the state office level, and there is no review by an independent third party outside of BLM state offices. In its review of the coal leasing procedures in 1984, the Linowes Commission concluded that periodic independent review of coal activities by a group with clear independence from the coal leasing program was desirable. Furthermore, both the *Uniform Standards of Professional Appraisal Practice* and the *Uniform Appraisal Standards for Federal Land Acquisition* note that independent appraisal review is an important tool for ensuring that the valuation estimate is credible. BLM headquarters officials currently have a very limited role in reviewing appraisal reports prior to a lease sale, and they told us that headquarters officials receive copies of between 5 and 10 percent of appraisal reports prior to a lease sale occurring. These officials told us that they are provided with these appraisal reports so that they can participate in sale panel meetings where BLM considers whether to accept bids for lease tracts. BLM headquarters officials do not sign off on these reports or provide comments to the state officials during the period when the appraisal reports are being developed. As a result of not regularly reviewing all appraisals, BLM headquarters officials were unaware of some of the differences in appraisal practices and documentation issues that we found across BLM state offices.

In addition, BLM is not currently taking advantage of a potential independent third-party reviewer with appraisal expertise within Interior, specifically, the Office of Valuation Services. The Office of Valuation Services, established by secretarial order in May 2010 and reorganized in Interior's Departmental Manual in June 2011, is responsible for providing real estate valuation services to the department's bureaus and offices, including "appraisals, appraisal reviews, consultation services, and mineral evaluation products for Department and client agencies."[51] Within the Office of Valuation Services, the Office of Mineral Evaluation is responsible for providing mineral evaluations for Interior's bureaus and offices, according to the *Departmental Manual.* Because the Office of Mineral Evaluation is a small office with about six staff, it is not feasible for this office to take over the mineral valuation function for the entire coal leasing program, according to officials in this office, and it would not be practical given the knowledge and expertise that state and field BLM staff have regarding coal in their respective regions. Rather, officials in this office said they were amenable to helping BLM in other ways by, for example, providing independent third-party review of appraisal reports, which is critical for ensuring the integrity of the appraisal process. Without additional oversight of the appraisal process by an independent reviewer, BLM is unable to ensure that its results are sound, key decisions are fully documented, and that differences we noted across state offices are warranted.

BLM Considers Coal Exports to Limited Extent When Estimating Fair Market Value and Does Not Consider Domestic Reserve Estimates Because of Their Variable Nature

BLM considers coal exports to a limited extent when developing an estimate of fair market value and generally does not explicitly consider estimates of the total amount of coal in the United States that can be mined economically, known as domestic reserve estimates. In the few state offices that did consider exports, we generally found the same generic statements in appraisal and economic reports that stated in general terms the possibility of future growth in coal exports, and there was limited tracking of exports from specific mines. As a result, BLM may not be factoring specific export information into appraisals or keeping up- to-date with emerging trends. Domestic reserve estimates are not considered due to the variable nature of these estimates according to BLM officials.

BLM Considers Coal Exports for Some Lease Sales in Certain States, but Associated Reports Provide Little Information on Current or Future Export Activity

BLM's guidance states that appraisal reports should consider specific markets for the coal being leased, and that "export potential" may be considered as part of the appraisal process. The export potential for coal from a particular mine can be influenced by several factors, including the quality of the coal and whether there is a transportation system nearby that can ship the large volume and weight of coal to a port for export.[52] Some coal mines, such as those in Wyoming's Powder River Basin, are part of a national coal market and, in 2011, Wyoming mines shipped coal to 34 states in the United States according to EIA data.[53] Other mines supply coal only to neighboring power plants, known as mine mouth operations, meaning that their export potential is limited, and exports would not factor into the fair market value estimation, according to BLM officials.

In our review of BLM case files for 31 coal lease sales, we found that coal exports were generally mentioned in appraisal and economic reports for the 13 federal lease sales held in Montana and Wyoming. Mines in these states exported 17.7 million tons of coal in 2011, according to EIA data, or about three-quarters of the total amount of coal exported from western states.[54] Exports from these states represented less than 2 percent of total U.S. coal production and about 17 percent of total U.S. exports of coal in 2011.[55] Of the 13 Montana and Wyoming case files we reviewed, one provided specific export information for the mine that was adjacent to the lease tract being appraised. This appraisal report, which was prepared for a lease tract in Montana, provided detailed information from IHS Global Insight and Wood Mackenzie, two private providers of information on coal.[56] In addition, we found that economic and appraisal reports in Wyoming typically contained generic boilerplate statements about the possibility of coal exports in the future and the uncertainty surrounding them, rather than specific information on actual or predicted coal exports—even for proposed lease tracts that were adjacent to mines on federal leases that are currently exporting coal. Wyoming BLM officials told us that coal exports made up such a small portion of total production from Wyoming that they did not believe it was necessary to provide specific information on exports in their economic or appraisal reports. Wyoming BLM officials told us that future appraisal reports may provide more specific export information if exports became a more significant issue, but they did not identify a threshold for including it.

We generally did not find mention of coal exports in the other states with federal coal leasing activity: Alabama, Colorado, Kentucky, New Mexico, North Dakota, Oklahoma, or Utah.[57] State BLM officials in these states told us they did not consider exports when estimating fair market value because there were few or no coal exports from their state. However, we found an example in Utah where the lease tract was adjacent to a mine that, according to EIA data, was exporting coal, but the appraisal report did not mention coal exports. EIA officials told us that they began collecting mine-level information on coal exports in 2008 and received a request from one BLM state office for these data. BLM state and headquarters officials generally told us they were not aware that EIA collects these data. Similarly, Wood Mackenzie has mine-level data on coal exports, but not all state BLM officials were aware that this information was available to them through a BLM subscription. By not tracking and considering all available export information, BLM may not be factoring specific export information into appraisals for lease tracts that are adjacent to mines currently exporting coal or keeping abreast of emerging trends in this area.

BLM officials said that they examine projections of future coal prices during the appraisal process, and these projections would account for exports. However, only the income approach for appraisals explicitly considers future prices, so the state offices that use only the comparable sales approach would not explicitly factor export potential into their fair market value assessments. Two states in particular—Colorado and Utah—have coal exports from mines on federal leases, but they generally use the comparable sales approach to estimate fair market value, therefore their fair market values would not explicitly reflect the potential impact of coal exports.

BLM officials told us that they are aware that some coal companies plan to export more coal in the future but voiced some concern about weighting these plans too heavily in estimating fair market value because major port infrastructure upgrades are needed on the West Coast to handle increased coal exports. Several stakeholders with expertise in coal markets that we interviewed shared this view. In addition, IEA said it is difficult to predict future coal exports from Wyoming's Powder River Basin to countries such as China because of a lack of infrastructure in place to handle exports and the uncertainty of market conditions.[58]

BLM Generally Does Not Consider Domestic Coal Reserve Estimates Because of Their Variable Nature

BLM officials told us that BLM does not consider domestic coal reserve estimates during the fair market value process.[59,60] One reason they gave was these estimates can vary greatly depending on market conditions. Domestic coal reserve estimates reflect the amount of coal that can be economically recovered at a given point in time; as a result, these estimates can change as coal prices fluctuate and mining technologies advance. For example, USGS estimated reserves of 10.1 billion tons in the Gillette coal field of the Powder River Basin at a sales price of $10.47 per ton in 2007,[61] but it changed this estimate to 18.5 billion tons when prices rose to $14.00 per ton in March 2008.[62] A more recent USGS assessment estimated that there was 25 billion tons of coal that can be economically recovered in the entire Powder River Basin at the time of study, but notes that "mining costs and coal prices are not static as both tend to increase over time."[63] The report goes on to state that "if market prices exceed mining costs, the reserve base will grow (the converse is also true)."

Some BLM officials told us they do not consider domestic reserve estimates when estimating fair market value because the United States has ample coal supplies to meet demand over the next 20 years, the time horizon that BLM uses when evaluating coal lease-by-applications. For example, EIA estimated that the United States has over 190 years of coal reserves, at the time of its most recent Annual Energy Outlook in April 2013.[64] While BLM does not consider reserve estimates explicitly, those BLM state offices that prepare an economic report as part of estimating fair market value examine future demand and price projections for coal, which impact reserve estimates as mentioned previously.

BLM Provides Limited Information on Federal Coal Lease Sales to the Public

BLM generally provides limited information on federal coal lease sales to the public. Environmental documents produced as part of the NEPA process and required coal lease sale announcements are the primary source of detailed written information made available on coal lease sales. The amount and type of information provided on websites vary by state office, with the most comprehensive information of the websites we reviewed provided by the Wyoming BLM state office. In addition, BLM does not typically make documents used to estimate fair market value publicly available due to the sensitive and proprietary information they contain, although its guidance states that a public version of the appraisal document should be prepared.

BLM Provides Some Information through Environmental Documents and Coal Lease Sale Announcements

BLM provides some information on coal lease sales in environmental documents developed to meet NEPA requirements and in lease sale announcements. BLM is required to share these documents with the public, and these documents are made available for review in public reading rooms in relevant BLM state and field offices and are also typically available on BLM's websites during the period of leasing activity. These environmental documents include environmental assessments and environmental impact statements,[65] which evaluate the likely environmental effects of leasing and mining the proposed lease tract. These documents generally include information on the lease applicant, mining methods at the existing operation, alternatives considered, and anticipated environmental effects. For example, an environmental assessment for a recent coal lease in Montana included an overview of the mine's history, the mining methods used at the site, the mine's layout, and information on potential effects of alternatives considered. In addition to environmental documents, a decision document summarizing the results of the process and the agency decision regarding the lease sale is also issued.[66]

BLM is also required to announce forthcoming coal lease sales in the *Federal Register* and a newspaper in the area of the lease tract. These announcements typically include general characteristics of the lease tract up for sale, such as the size of the tract, and the amount and quality of the coal being offered, including its estimated heating value, ash and moisture content, and the thickness of the coal beds. In addition, the announcements list the applicant and potential use of the tract, such as whether it will be used to extend existing mining operations or the tract's location adjacent to more than one existing mine. The announcement also notes where interested stakeholders can view lease sale details including bidding instructions, terms and conditions of the proposed coal lease, and case file documents, typically available for review at the relevant BLM state office.

BLM State Offices Vary in the Amount and Type of Coal Lease Information on Their Websites

BLM websites are another way that public information is released on the leasing program, but we found that it was difficult to locate this information on some of BLM's websites that we reviewed and the amount and type of information shared across the websites that we accessed in May 2013 varied (see table 3). For example, BLM headquarters' website contains general information on the federal coal leasing program, but it does not include information on past or upcoming federal coal lease sales or link to relevant BLM state or field office websites. BLM officials told us that they attempted to provide general information on past lease sales on the headquarters website in 2010, but they were unable to obtain state BLM offices' verification of the data, which stalled the effort. Five of the six state offices do not maintain information on past lease sales on their websites, although officials in BLM headquarters and two state offices also told us they have provided this information upon request. All six state offices that manage lease sales, at a minimum, publish lease sale announcements in the *Federal Register*, which is searchable via the Internet, and based on our review of BLM websites and interviews with BLM officials, all but one of the state offices issue press releases with lease sale results that are highlighted for limited periods. In addition, during our review of BLM websites, we found that five of the six state offices keep environmental documents related to lease sales on their websites during the time of lease sale activity.

Table 3. Summary of Federal Coal Leasing Information Contained on BLM Websites

BLM office website and states covered	Colorado BLM (Colorado)	Eastern States BLM (Alabama, Kentucky)	Montana/ Dakotas BLM (Montana, North Dakota)	New Mexico BLM (New Mexico, Oklahoma)	Utah BLM (Utah)	Wyoming BLM (Wyoming)
Number of coal tracts leased since 1990	20	13	16	12	15	31
General information on the federal coal leasing program[a]	—	√	√	—	√	√
Lease sale results announced in press release	√	√	√	—	√	√
Environmental documents linked on website (during time of lease sale activity)	√	√	√	—	√	√
Final environmental documents maintained on website	—	—	—	—	—	√
Summary information from past lease sales maintained on website	—	—	—	—	—	√

Sources: GAO analysis of BLM websites accessed May 2013 and interviews with BLM officials.

[a] Links to general information on primary coal website pages were considered as being contained on BLM website pages.

Of the six state office websites we reviewed, the Wyoming state office provided the most comprehensive information on the federal coal leasing program, including results for all coal lease sales in the Powder River Basin since 1990. For each lease sale, this website had information on successful bid amounts, associated coal volume and coal quality, and links to environmental documents. Wyoming BLM officials told us that they had this information on their website because they receive regular inquiries from the press and public on coal leasing in the Powder River Basin. In contrast, the New Mexico state office had no coal leasing information on its website. New Mexico BLM officials told us that there is not much public interest in coal lease sales in the states of New Mexico and Oklahoma, which they oversee, and requests for this type of information are limited to inquiries from mining companies.

Making electronic information available to the public is a position supported by the Office of Management and Budget (OMB) and has been demonstrated by other agencies. Specifically, OMB guidance directs federal agencies to use electronic media to make government information more easily accessible and useful to the public.[67] In addition, we have previously reported on the importance of federal programs allowing users to easily access and use information on websites.[68] BLM's federal oil and gas onshore leasing program maintains a list of planned lease sale auction dates on the headquarters level website, along with summary results from recent lease sales by state. Without standard information on BLM websites, federal coal leasing activity is difficult to track by the public, and access to publicly available documents may be hampered.

BLM Does Not Make Reports Related to Its Estimation of Fair Market Value Publicly Available, Which Is Inconsistent with Some Parts of Its Guidance

BLM's guidance states that a public version of the appraisal report that deletes all proprietary material should be prepared for each lease sale, but BLM has not been following this guidance.[69] According to officials from BLM state offices, a public version of appraisal reports is not prepared as a standard practice in the six BLM offices managing the coal lease sale process. According to some BLM officials, they do not prepare this public version because they are concerned about the potential release of proprietary and sensitive information these reports contain and the impact this could have on the bidding process.[70]

BLM's guidance also states that the "fair market value appraisals and estimates can be released to the public upon request on tracts where the high bid has been accepted," and further states "information and analyses documents used to derive these released fair market value estimates are to be released to the public upon request" after these documents have been "modified to exclude proprietary information."[71] BLM has interpreted this guidance to mean that the agency has the discretion to determine whether to release these reports in a redacted format. For two Freedom of Information Act (FOIA) requests received in 2011 for reports used to determine fair market value of coal leases,[72] BLM initially withheld all fair market value documents until Interior's Office of the Solicitor advised BLM to provide redacted documents in response to an appeal filed in one of these cases.[73] In its response to this FOIA appeal, Interior's Office of the Solicitor agreed that BLM has discretionary authority to disclose this information and noted that BLM's guidance "does not require the BLM to release 'fair market value appraisals and estimates' to the public and, instead, merely notes that it 'can' do so." In the end, BLM provided redacted appraisal reports to this FOIA request, which we reviewed. These documents included a description of the approaches BLM used to estimate fair market value, the number of comparable sales that were considered, and background information on the mining operation, but the fair market value estimate was redacted along with the supporting analysis behind this number. As of June 2013, BLM was in the process of responding to another request for fair market value documents received in 2012.

BLM headquarters and state office officials consistently told us that it is critical that the sensitive information in lease sale documents not be released publicly so that the integrity of the sealed bid process can be maintained. For example, if companies were to obtain the specific comparable sales used for a past lease sale, this information could lead them to reduce their bid for a future lease sale so that it is closer to the fair market value estimate, according to BLM officials. But there are differing views within the agency on the extent of information that should and could be shared. For instance, BLM headquarters officials told us that they are open to releasing additional information on federal coal leasing, including making redacted appraisal reports available. In contrast, Wyoming BLM officials told us they were not comfortable making any additional information on the fair market value process available such as redacted appraisal reports. They told us that, in their opinion, considerable information is already available in documents that must be prepared as part of the process, such as environmental impact statements, public notices, and detailed statements on how to bid. They also told us most people are interested primarily in lease sale results, which Wyoming BLM makes available on its website. Wyoming BLM officials also said they are concerned that, by making additional information available, including redacted appraisal reports, some important information might be shared that would result in reduced bids on future coal lease sales. The Wyoming BLM officials' point of view stands in conflict with BLM's guidance that additional information in the form of public versions of the appraisal report should be prepared and the Office of the Solicitor's determination that FOIA does not allow BLM to withhold entire documents relating to the estimate of fair market value in response to FOIA requests when portions of these documents contain information that is not protected from disclosure and should be released.

CONCLUSION

With about 40 percent of the nation's coal produced from federal coal lease tracts in recent years, the federal coal leasing program plays an important role in the nation's energy portfolio. In managing the leasing program, BLM is required to obtain fair market value for coal leases. Because there is typically little competition for federal leases, BLM plays a critical role in ensuring that the public receives fair market value for the coal that is leased. However, we found differences across BLM state offices in the approaches they use to estimate fair market value and the rigor of these reports. Moreover, BLM state offices are not documenting the rationale for choosing their approaches for the appraisal process.

Adequate oversight of the fair market value process is critical to ensuring that its results are sound and properly reviewed. However, BLM's guidance on the valuation of coal properties is out of date, and officials are not reviewing and signing appraisal reports in accordance with BLM's guidance. Without a mechanism to ensure consistent reviews by three officials, as specified in the guidance, and independent third-party reviews, appraisal reports may not be receiving the scrutiny they deserve.

BLM's guidance allows for additional information and analyses to be considered as part of the postsale review process, which could result in a lower revised fair market value estimate and acceptance of bids below the presale fair market value estimate but above the revised estimate. The guidance calls for such decisions to be fully justified and that a revised fair market value be clearly documented and reviewed. However, we found instances where BLM's justification to accept such bids was not adequately documented. Without proper documentation of these decisions, adequate oversight cannot take place, and BLM does not have assurance that accepted bids were in compliance with the Minerals Leasing Act.

Coal exports make up a small but growing proportion of total U.S. coal production, yet BLM state offices were generally not tracking the export activity for mines on federal leases and were including only generic statements about exports in their appraisal reports, and some state offices were not routinely including export information in appraisal reports. Moreover, BLM officials were largely unaware of the various sources of mine-level information about exports, such as the information that EIA collects and the information collected by private companies. By not tracking and considering all available export information, BLM may not be factoring specific export information into appraisals for lease tracts that are adjacent to mines currently exporting coal or keeping abreast of emerging trends in this area.

BLM state offices are not following agency guidance because they have not prepared public versions of appraisal reports, and there is a lack of agreement within the agency on the extent and type of information related to the estimation of fair market value to be shared in response to public requests. Without updated guidance and a consensus, there may continue to be a disconnect between BLM's guidance and its standard practice of not releasing this information publicly. Finally, BLM provides little summary information on its websites on past lease sales or links to sale-related documents. Having additional information online could increase the transparency of federal coal leasing program.

RECOMMENDATIONS FOR EXECUTIVE ACTION

We are recommending that the Secretary of the Interior direct the Director of the Bureau of Land Management to take the following eight actions:

To ensure that appraisal reports reflect future trends in coal markets, BLM should revise its guidance to have state offices use both comparable sales and income approaches to estimate fair market value where practicable. Where it is not practicable to do so, the rationale should be documented in the appraisal report.

To ensure that appraisal reports receive the scrutiny they deserve and are reviewed by specified officials, BLM should take the following actions:

- update its guidance so that it reflects the current titles of officials who should review appraisal reports;
- develop a mechanism to ensure that state offices are reviewing and signing appraisal reports consistent with the guidance;

- develop a process for independent review of appraisal reports and work with the Office of Valuation Service to determine its role, if any, in this process.

To ensure that all accepted bids comply with the Minerals Leasing Act by meeting or exceeding BLM's estimate of fair market value, BLM should update its guidance to specify the documentation needed for postsale analyses in instances where a decision is made to revise the fair market value estimate and accept a bonus bid that was below the presale estimate of fair market value but above the revised estimate. Such documentation for postsale analyses should include the revised estimate of fair market value, the rationale for this revision, and review of this decision by appropriate officials.

To ensure that appraisal reports reflect the current state of export activity for mines on federal leases, BLM headquarters should develop guidance on how to consider exports as part of the appraisal process and identify potential sources of information on coal exports that state offices should use when conducting appraisals.

To eliminate the disconnect between its guidance and BLM state offices' practice of not releasing appraisal documents to the public, BLM headquarters, state office officials, and Interior's Office of the Solicitor should come to agreement on the extent and type of information related to the estimation of fair market value that should be shared in response to public requests for this information and make sure that its guidance reflects this consensus.

To make electronic information on the coal leasing program more accessible to the public, BLM should provide summary information on its websites on results of past lease sales (e.g., amount of coal offered, coal quality, bonus bids received) and status of any upcoming coal lease sales along with links to sale-related documents.

AGENCY COMMENTS

We provided a draft of this report to the Department of the Interior, the Department of Agriculture, and the Department of Energy for review and comment. The Department of the Interior concurred with our recommendations and also noted it has begun to address some of these recommendations. Specifically, BLM has signed a memorandum of understanding with the Office of Valuation Services to enhance the review of fair market values. In addition, BLM stated it will soon publish additional information on lease sales on its national and state websites. The Departments of the Interior, Agriculture, and Energy also provided us with technical comments, which we have incorporated as appropriate.

Anne-Marie Fennell
Director, Natural Resources and Environment

APPENDIX I: OBJECTIVES, SCOPE AND METHODOLOGY

Our objectives were to examine (1) federal coal leasing, including the number of tracts leased, along with the trends in associated coal production and revenues generated since 1990; (2) Bureau of Land Management's (BLM) implementation of the process to develop an estimate of fair market value for coal leases; (3) the extent to which BLM considers coal exports and domestic coal reserve estimates when developing an estimate of fair market value; and (4) the extent to which BLM communicates information on federal coal lease sales to the public.

To provide information on trends in federal coal leasing under the first objective, we analyzed data from BLM's LR2000 database—used by BLM to track federal land and mineral resources including coal—and summarized federal coal lease sale activity and bonus bids accepted from January 1, 1990 to December 31, 2012. For each lease sale where a bid was accepted and the tract leased, we analyzed data including: lease sale date, tract acreage, the amount of offered coal, number of bids received, and winning bid amounts. We also analyzed data on coal production and revenues generated from federal coal leases from fiscal years 1990 to 2012 from the Department of the Interior's Office of Natural Resources and Revenue (ONRR), which is responsible for collecting and distributing revenues associated with federal mineral leases including federal coal leases. We used ONRR sales year revenue data, which includes current fiscal year data and adjusted or corrected transactions for sales that took place in previous years. According to ONRR officials, adjustments to sales year data are made on an ongoing basis in real time, such that the data varies daily. We used sales year data because this type of data was identified by ONRR as the best for trending purposes. To complete our analysis, we adjusted both BLM bonus bid data and ONRR revenue data to 2013 dollars using the gross domestic product price index.

We conducted interviews with BLM and ONRR officials regarding these data and reviewed documentation on their data systems. We found that some of the revenue data initially provided by ONRR prior to 2003, in particular the bonus, rent, and other income data, had gaps resulting from a data system conversion the agency underwent and was not reliable for use in our analysis. ONRR ultimately provided updated bonus data for this period, but it did so late in our review process, and we were unable to determine its reliability. We determined that all other ONRR data including royalty and production data from 1990 to 2012, as well as BLM federal coal leasing data, were sufficiently reliable for describing trends in the federal coal leasing program.

To examine how BLM implements the process to develop an estimate of fair market value, we reviewed applicable regulations and BLM's guidance for the coal leasing program, including BLM's H-3070-1 handbook, titled *Economic Evaluation of Coal Properties*. We also interviewed BLM officials in headquarters and state offices on how they implement these regulations and guidance. Specifically, we interviewed officials in the following BLM state offices because they are the only state offices involved in federal coal leasing at BLM: Colorado, Eastern States, Montana/Dakotas, New Mexico, Utah, and Wyoming.[74] We also spoke with officials in the Casper Field Office who are directly involved in coal leasing activity in the Powder River Basin. In addition, we reviewed other appraisal standards developed by appraisal organizations in the United States and appraisal standards used in other countries. These standards included the *Uniform Standards of Professional Appraisal Practice* prepared by the Appraisal Standards Board in the United States; the *Uniform Appraisal Standards for Federal Lands Acquisitions* prepared by the Interagency Land Acquisition Conference in the United States; *Standards and Guidelines for Valuation of Mineral Properties* prepared by the Canadian Institute of Mining, Metallurgy and Petroleum; and the *Code for Technical Assessment and Valuation of Mineral and Petroleum Assets and Securities for Independent Expert Reports* prepared by several groups, including the Australasian Institute of Mining and Metallurgy. We examined these standards to see what they said about certain aspects of an appraisal including required documentation and review processes. To learn about appraisal practices for mineral properties, we also spoke with appraisal officials, including officials from the Appraisal Institute, the Appraisal Foundation, the American Institute of Mineral Appraisers, and an official involved in the development of the Canadian standards for mineral valuation mentioned above. In addition, we spoke with officials from Interior's Office of Valuation Services, which is responsible for providing real estate evaluation services to the Department of the Interior's bureaus and offices.

We selected and reviewed a nonrandom sample of case files prepared by BLM officials as part of 31 recent coal lease sales using a data collection instrument we developed. The sample included all reports for lease sales that generally took place from January 1, 2007, to July 31, 2012. This nonrandom sample cannot be generalized to all coal lease sales held but rather has a focus on recently prepared files. However, the results of this sample provide illustrative examples of the coal leasing process used and the documentation prepared. We requested the following documentation from BLM for these lease sales if they had been prepared: appraisal report, economic report, engineering report, geologic report, and tract modification report. As part of our review, we examined 147 documents that were prepared for these 31 lease sales. For those states that did not oversee two lease sales from January 1, 2007, to July 31, 2012— Alabama, Kentucky, New Mexico, North Dakota, and Oklahoma—we examined their two most recent lease sales.[75] In the end we reviewed case files for 11 lease sales in Wyoming, 4 lease sales in Colorado, 3 lease sales in both Oklahoma and New Mexico, and 2 lease sales each in Alabama, Kentucky, North Dakota, Montana, and Utah.[76] To ensure that our data collection instrument was filled out correctly, two GAO staff members reviewed the provided documents: one filled out the data collection instrument the first time, and the other verified this work. We conducted follow-up interviews with BLM state offices to discuss both general questions our review raised about the processes used to estimate fair market value in each of the BLM states and details related to specific cases we reviewed.

To determine the extent to which BLM considers coal exports when developing an estimate of fair market value, we used our case file review to examine what types of information BLM provided on exports, if any. For those files that did contain information on exports, we compared the wording used to describe exports across the various reports to see what kind of information was provided. We also used our interviews with BLM officials at headquarters and state offices to learn about the information they consult in estimating fair market value and the extent to which they consider exports when making this determination. We examined available information on coal exports from the Energy Information Administration (EIA) and other publicly available documents, such as financial statements of mining companies. We also spoke with knowledgeable stakeholders about future projections for coal exports, including the National Mining Association, International Energy Agency, and other officials from academia and industry.

To determine the extent to which BLM considers reserve estimates, we interviewed a variety of BLM officials at the headquarters and state office level to determine if reserves were considered. In addition, we examined available reserve information from the United States Geological Survey (USGS) and spoke with USGS officials involved in making these estimates. We also obtained perspectives from stakeholders from academia, industry, and environmental organizations.

To examine the extent to which BLM provides information to the public on coal lease sales, we analyzed BLM's policies for making information publicly available, including BLM's H-3070-1 handbook. We also reviewed BLM websites related to federal coal leasing, and we reviewed a sample of environmental documents that are made publicly available during the coal leasing process. We obtained data from BLM on Freedom of Information Act (FOIA) requests made for fair market value information prepared for federal coal lease sales. We also reviewed copies of request letters and BLM's response to these requests, including redacted versions of fair market value documents made available in response to the only FOIA request where BLM supplied these documents. We interviewed BLM staff, industry representatives, as well as conservation and environmental groups to get their perspectives on the information made publicly available on federal coal leases.

Finally, we conducted site visits to Colorado and Wyoming. During these visits, we met with officials in BLM state offices in Colorado and Wyoming, and we also met with officials in the Casper Field Office in Wyoming. In addition, we met with a coal mining company and toured a large surface mine in Wyoming and met with a professor of economics at the University of Wyoming's School of Energy Resources. We selected these states because they have different types of mining that take place—generally surface mining in Wyoming and underground mining in Colorado. In addition, we selected Wyoming because of the large amount of federal coal leasing activity in the state.

We conducted this performance audit from June 2012 to December 2013 in accordance with generally accepted government auditing standards. Those standards require that we plan and perform the audit to obtain sufficient, appropriate evidence to provide a reasonable basis for our findings and conclusions based on our audit objectives. We believe that the evidence obtained provides a reasonable basis for our findings and conclusions based on our audit objectives.

APPENDIX II: FEDERAL COAL LEASE SALES, JANUARY 1990 THROUGH DECEMBER 2012

This appendix presents data on all federal coal lease sales by state that were conducted from January 1, 1990, through December 31, 2012. Table 4 provides information on the lease tract characteristics (acreage, type of mine, and amount of coal) along with the lease sale results (number of bids received, bonus bid accepted, and name of successful bidder).

APPENDIX III: SUMMARY INFORMATION FROM FILE REVIEWS OF SELECTED FEDERAL COAL LEASE SALES

This appendix provides information on the 31 federal coal lease sales we reviewed that generally took place from January 1, 2007, to July 31, 2012. For those BLM state offices that did not conduct 2 lease sales during this time, we reviewed their 2 most recent lease sales.[77]

Reports that are relevant to the determination of fair market value include the following:

- geologic reports, which contain an estimate of the amount of coal that can be recovered on the lease tract along with the characteristics of the coal, including its heating content;
- engineering reports, which generally contain estimates of the costs to extract the coal based on the number of employees and capital equipment necessary to carry out mining activities;
- economic reports, which establish price and demand levels for the lease tract's coal; and
- appraisal reports, which document the fair market value for the lease tract, along with an explanation of the methods used to develop this number.

BLM's guidance does not direct that all of these reports to be prepared as part of a lease sale. For example, it is unlikely that an economic report would be prepared if the income approach was not used to determine fair market value. However, BLM guidance requires that appraisal reports be signed by three officials.

Table 4. Federal Coal Lease Sales, January 1990 through December 2012

Sale date	Lease serial number	Acres	Associated mine name	Type of mine	Estimated amount of coal leased (1,000 tons)[a]	Number Of qualified bids received	Total accepted bonus bid (nominal dollars)	Bonus bid per Ton (nominal dollars)	Bonus bid per Acre (nominal dollars)	Successful bidder
Alabama										
6/12/1990	ALES 041886	80	River King	Surface	160	1	$21,101	$0.13	$264	River King Energy
7/12/1991	ALES 043165	6440	Yellow Creek	Underground	24,600	1	$1,300,880	$0.05	$202	Pittsburg & Midway Mining Coal Co.
3/18/1994	ALES 044853	1610	Shoal Creek	Underground	8,065	1	$330,000	$0.04	$205	Drummond Co. Inc.
6/28/1996	ALES 046611	40	Mary Lee Number 1	Underground	191	1	$7,795	$0.04	$194	Drummond Co. Inc.
8/21/1997	ALES 047886	40	Oak Mountain and Boone Number 1	Underground	500	1	$4,223	$0.00001	$103	Oak Mtn. Energy LLC
9/30/2004	ALES 051589	2887	North River	Underground	10,789	1	$320,568	$0.03	$111	Pittsburg & Midway Mining Coal Co.
Colorado										
4/27/1990	COC 049465	193	King	Underground	1,000	2	$101,560	$0.10	$526	National King Coal

Table 4. (Continued)

Sale date	Lease serial number	Acres	Associated mine name	Type of mine	Estimated amount of coal leased (1,000 tons)[a]	Number Of qualified bids received	Total accepted bonus bid (nominal dollars)	Bonus bid per Ton (nominal dollars)	Bonus bid per Acre (nominal dollars)	Successful bidder
6/26/1991	COC 051551	1280	Deserado	Underground	8,700	1	$656,640	$0.08	$513	Western Fuels
7/30/1992	COC 053510	1340	Elk Creek	Underground	10,300	1	$1,025,998	$0.10	$766	Somerset Mining
12/18/1992	COC 053356	522	Bowie Number 2	Underground	2,500	1	$52,178	$0.02	$100	Cyprus Orchard Valley
10/7/1993	COC 053560	544	Foidel Creek	Underground	3,600	1	$57,225	$0.02	$105	Cyprus Western
11/4/1993	COC 054558	1012	West Elk	Underground	14,000	1	$803,520	$0.06	$794	Mountain Coal Co. LLC
5/15/1995	COC 056447	2770	West Elk	Underground	37,000	1	$6,408,035	$0.17	$2,314	Mountain Coal Co. LLC
1/25/1996	COC 054608	2600	Foidel Creek	Underground	23,870	1	$4,057,900	$0.17	$1,561	Twenty mile Coal Co.
3/30/1998	COC 060941	195	King	Underground	624	1	$19,752	$0.03	$101	National King Coal
5/31/2000	COC 061357	4444	Elk Creek	Underground	20,920	1	$8,723,661	$0.42	$1,963	Oxbow Mining Inc.
5/31/2000	COC 061209	3211	Bowie Number 2	Underground	32,600	1	$10,334,186	$0.32	$3,219	Bowie Re-sources Ltd.

Sale date	Lease serial number	Acres	Associated mine name	Type of mine	Estimated amount of coal leased (1,000 tons)[a]	Number Of qualified bids received	Total accepted bonus bid (nominal dollars)	Bonus bid per Ton (nominal dollars)	Bonus bid per Acre (nominal dollars)	Successful bidder
12/4/2001	COC 062920	1305	King II	Underground	7,049	1	$396,720	$0.06	$304	National King Coal
1/6/2004	COC 066514	1520	McClane Canyon	Underground	8,365	2	$2,065,000	$0.25	$1,358	Central Appalachia Mining, CAM Holdings
7/1/2004	COC 067011	691	West Elk	Underground	2,300	1	$191,994	$0.08	$278	Arch Coal, Inc., Ark Land Co.
10/12/2006	COC 067514	200	Foidel Creek	Underground	2,100	1	$525,000	$0.25	$2,620	Twentymile Coal Co.
1/24/2007	COC 067232	1517	West Elk	Underground	14,000	1	$3,025,000	$0.22	$1,994	Ark Land Co.
5/30/2007	COC 068590	1407	Colowyo	Surface	92,000	1	$13,106,600	$0.14	$9,317	Colowyo Coal Co. LP
1/14/2009	COC 072980	500	Foidel Creek	Underground	1,400	1	$350,000	$0.25	$700	Twentymile Coal Co.
5/15/2012	COC 070615	726	Elk Creek	Underground	3,960	1	$990,264	$0.25	$1,364	Oxbow Mining, LLC
8/22/2012	COC 074219	400	Sage Creek	Underground	3,200	1	$800,000	$0.25	$2,000	Sage Creek Holdings, LLC
Kentucky										
12/13/1990	KYES 041395	180	Bell Co. Number 2	Underground	900	3	$196,000	$0.22	$1,089	Apollo Fuel

Table 4. (Continued)

Sale date	Lease serial number	Acres	Associated mine name	Type of mine	Estimated amount of coal leased (1,000 tons)[a]	Number Of qualified bids received	Total accepted bonus bid (nominal dollars)	Bonus bid per Ton (nominal dollars)	Bonus bid per Acre (nominal dollars)	Successful bidder
12/13/1990	KYES 045303	800	Stone Number2	Underground	3,800	1	$160,000	$0.04	$200	Apollo Fuel
5/3/1991	KYES 043034	167	Camp Number 2	Underground	900	1	$16,658	$0.02	$100	Peabody
5/3/1991	KYES 042948	99	Big Fist Number 4	Underground	400	1	$9,900	$0.02	$100	Wellmore Coal
8/31/1993	KYES 045088	181	Number 60	Underground	500	1	$19,005	$0.04	$105	Leeco
7/27/2005	KYES 051002	1210	Beechfork	Underground	2,900	1	$501,003	$0.17	$414	Bledsoe Coal Leasing Company
5/11/2006	KYES 050213	315	Chas Number 4	Underground	792	1	$47,250	$0.06	$150	Chas Coal LLC
Montana										
6/16/1999	MTM 080697	1401	Rosebud	Surface	27,600	1	$4,416,000	$0.16	$3,152	Western Energy Co.
11/27/2000	MTM 088405	150	Spring Creek	Surface	15,400	1	$1,740,200	$0.11	$11,601	Spring Creek Coal Co.
4/17/2007	MTM 094378	1118	Spring Creek	Surface	108,600	1	$19,902,200	$0.18	$17,806	Spring Creek Coal Co.
2/28/2012	MTM 097988	2680	Bull Mountains Mine No. 1	Surface	35,500	1	$10,650,000	$0.30	$3,974	Signal Peak Energy LLC

Sale date	Lease serial number	Acres	Associated mine name	Type of mine	Estimated amount of coal leased (1,000 tons)[a]	Number Of qualified bids received	Total accepted bonus bid (nominal dollars)	Bonus bid per Ton (nominal dollars)	Bonus bid per Acre (nominal dollars)	Successful bidder
New Mexico										
7/31/1991	NMNM 078371	3082	Salt River	Surface	19,600	1	$1,571,963	$0.08	$510	Salt River Project
7/31/1991	NMNM 086717	3360	Salt River	Surface	10,000	1	$799,680	$0.08	$238	Salt River Project
11/1/2000	NMNM 099144	4484	San Juan	Underground	63,000	1	$13,000,000	$0.21	$2,899	San Juan Coal Co.
North Dakota										
1/23/1991	NDM 078697	160	Freedom	Surface	2,500	1	$16,000	$0.01	$100	Coteau Properties Co.
8/9/1994	NDM 081582	793	Freedom	Surface	9,140	1	$79,300	$0.01	$100	Coteau Properties Co.
3/26/1997	NDM 085516	159	Freedom	Surface	1,750	1	$15,900	$0.01	$100	Falkirk Mining Co.
3/26/1997	NDM 085537	80	Coteau	Surface	510	1	$8,000	$0.02	$100	Coteau Properties Co.
3/26/1997	NDM 085517	399	Freedom	Surface	5,610	1	$39,900	$0.01	$100	Coteau Properties Co.

Table 4. (Continued)

Sale date	Lease serial number	Acres	Associated mine name	Type of mine	Estimated amount of coal leased (1,000 tons)[a]	Number Of qualified bids received	Total accepted bonus bid (nominal dollars)	Bonus bid per Ton (nominal dollars)	Bonus bid per Acre (nominal dollars)	Successful bidder
3/26/1997	NDM 085515	79	Freedom	Surface	2,000	1	$8,000	$0.004	$101	Coteau Properties Co.
9/30/1998	NDM 086601	360	Beulah	Surface	6,210	1	$36,000	$0.01	$100	Knife River Corp.
2/12/2002	NDM 090783	503	Freedom	Surface	7,000	1	$50,280	$0.01	$100	Coteau Properties Co.
12/10/2002	NDM 091647	40	Falkirk	Surface	300	1	$4,000	$0.01	$100	Falkirk Mining Co.
3/2/2006	NDM 091535	5334	Freedom	Surface	89,000	1	$533,400	$0.01	$100	Coteau Properties Co.
9/12/2006	NDM 095104	320	Center	Surface	8,300	1	$32,000	$0.004	$100	BNI Coal LTD
10/15/2009	NDM 097633	160	Center	Surface	3,000	1	$16,000	$0.01	$100	BNI Coal LTD
Oklahoma										
11/14/1994	OKNM 091590	400	Shady Point, Cavanal	Underground	369	1	$165,544	$0.45	$414	Farrell-Cooper Mining Co.
11/14/1994	OKNM 091569	2725	Heavener	Surface	18,290	1	$281,388	$0.02	$103	Farrell-Cooper Mining Co.

Sale date	Lease serial number	Acres	Associated mine name	Type of mine	Estimated amount of coal leased (1,000 tons)[a]	Number Of qualified bids received	Total accepted bonus bid (nominal dollars)	Bonus bid per Ton (nominal dollars)	Bonus bid per Acre (nominal dollars)	Successful bidder
11/14/1994	OKNM 091568	1933	Red Oak South	Surface	11,137	1	$204,502	$0.02	$106	Farrell-Cooper Mining Co.
11/14/1994	OKNM 091190	3429	Pollyanna Number 8 North and South	Surface	15,320	1	$346,710	$0.02	$101	P&K Company
1/31/1997	OKNM 094663	90	Pollyanna Number 9	Surface	138	1	$10,000	$0.07	$111	Georges Colliers Inc.
1/31/1997	OKNM 091571	2120	Rock Island	Surface	4,300	1	$251,816	$0.06	$119	Farrell-Cooper Mining Co.
9/14/2005	OKNM 108097	2380	McCurtain	Underground	10,058	2	$323,204	$0.03	$136	Farrell-Cooper Mining Co.
9/14/2005	OKNM 107920	2702	Bull Hill	Surface	8,993	2	$409,965	$0.05	$152	Farrell-Cooper Mining Co.
9/14/2005	OKNM 104763	560	Liberty Number 4	Surface	2,057	1	$57,820	$0.03	$103	Farrell-Cooper Mining Co.
Utah										
1/11/1990	UTU 064263	1987	Star Point Mine	Underground	7,630	1	$1,526,369	$0.20	$768	Cyprus Western
6/28/1990	UTU 066060	933	Aberdeen	Underground	8,800	1	$1,654,776	$0.19	$1,773	AMCA Coal Leasing

Table 4. (Continued)

Sale date	Lease serial number	Acres	Associated mine name	Type of mine	Estimated amount of coal leased (1,000 tons)[a]	Number Of qualified bids received	Total accepted bonus bid (nominal dollars)	Bonus bid per Ton (nominal dollars)	Bonus bid per Acre (nominal dollars)	Successful bidder
6/28/1990	UTU 064375	2631	Trail Mountain	Underground	12,200	2	$6,103,479	$0.50	$2,320	Beaver Creek Coal
12/29/1993	UTU 068082	2979	Genwal, Crandall Canyon North	Underground	18,600	1	$3,810,000	$0.20	$1,279	Andalex Resources, Inc.
7/19/1995	UTU 069635	2177	Dugout Canyon	Underground	12,700	1	$2,667,000	$0.21	$1,225	Sage Point Coal Co.
5/30/1996	UTU 067939	3291	Skyline	Underground	24,100	1	$5,600,000	$0.23	$1,702	Coastal States Energy Co.
12/18/1996	UTU 073975	2299	Willow Creek Mine	Underground	22,100	1	$5,127,200	$0.23	$2,230	Cyprus Plateau Mining Co.
5/14/1998	UTU 074804	1288	Horizon	Underground	6,300	1	$315,000	$0.05	$244	White Oak-Horizon Mining
5/20/1999	UTU 076195	7172	Sufco	Underground	60,000	1	$16,900,000	$0.28	$2,356	Canyon Fuel Co.
12/12/2001	UTU 078562	1646	West Ridge	Underground	14,800	1	$11,459,900	$0.77	$6,961	Andalex Resources, Inc.
6/12/2003	UTU 078953	880	Genwal, Crandell Canyon South	Underground	7,630	1	$6,561,800	$0.86	$7,457	Andalex Resources, Inc.

Sale date	Lease serial number	Acres	Associated mine name	Type of mine	Estimated amount of coal leased (1,000 tons)[a]	Number Of qualified bids received	Total accepted bonus bid (nominal dollars)	Bonus bid per Ton (nominal dollars)	Bonus bid per Acre (nominal dollars)	Successful bidder
6/24/2004	UTU 079975	703	Aberdeen	Underground	3,040	1	$486,400	$0.16	$692	Andalex Resources, Inc.
6/8/2006	UTU 081893	1760	Aberdeen	Underground	14,900	1	$2,816,000	$0.19	$1,600	Andalex Resources, Inc.
8/1/2006	UTU 084285	214	Deer Creek	Underground	325	1	$98,000	$0.30	$459	PACIFICORP
9/3/2009	UTU 086038	120	Emery Deep	Underground	561	1	$201,600	$0.36	$1,680	Consolidation Coal
Wyoming										
9/20/1990	WYW 119606	81	Swanson	Underground	300	1	$28,611	$0.10	$353	Lion Coal
9/26/1991	WYW 117924	1709	Jacobs Ranch	Surface	102,600	1	$20,110,457	$0.20	$11,770	Kerr McGee
8/12/1992	WYW 118907	3493	Black Thunder	Surface	481,000	1	$71,898,842	$0.15	$20,587	Thunder Basin
9/28/1992	WYW 119554	3064	North Antelope Rochelle	Surface	370,000	1	$86,986,960	$0.24	$28,390	Powder River
1/7/1993	WYW 122586	463	Rocky Butte	Surface	55,000	1	$16,500,001	$0.30	$35,621	Northwestern Resources
4/5/1995	WYW 124783	1059	Eagle Butte	Surface	166,400	1	$18,470,400	$0.11	$17,438	AMAX Land Co.
12/4/1996	WYW 128322	617	Antelope	Surface	60,364	1	$9,054,600	$0.15	$14,670	Antelope Coal Co.

Table 4. (Continued)

Sale date	Lease serial number	Acres	Associated mine name	Type of mine	Estimated amount of coal leased (1,000 tons)[a]	Number Of qualified bids received	Total accepted bonus bid (nominal dollars)	Bonus bid per Ton (nominal dollars)	Bonus bid per Acre (nominal dollars)	Successful bidder
9/25/1997	WYW 127221	1482	North Rochelle	Surface	157,610	1	$30,576,340	$0.19	$20,633	Triton
6/30/1998	WYW 136142	4224	North Antelope Rochelle	Surface	532,000	1	$109,596,500	$0.21	$25,945	Powder River Coal Co.
10/1/1998	WYW 136458	3546	Black Thunder	Surface	412,000	2	$158,000,009	$0.38	$44,563	Arch Coal
9/29/1999	WYW 139975	5206	Elk Mountain	Surface & Underground	65,780	1	$1,957,456	$0.03	$376	Ark Land Co.
9/7/2000	WYW 141435	2819	Antelope	Surface	275,577	1	$91,220,121	$0.33	$32,363	Kennecott Energy
1/16/2002	WYW 146744	4982	Jacobs Ranch	Surface	537,542	2	$379,504,652	$0.71	$76,171	Kennecott Energy
6/29/2004	WYW 154001	2957	North Antelope Rochelle	Surface	297,469	1	$274,117,684	$0.92	$92,710	Peabody-BTU Western Resources, Inc.
9/22/2004	WYW 150318	5084	Black Thunder	Surface	718,719	1	$610,999,950	$0.85	$120,193	Ark Land LT Incorporated
11/17/2004	WYW 151634	921	Buckskin	Surface	142,698	1	$42,809,400	$0.30	$46,473	Kiewit Mining Group

Sale date	Lease serial number	Acres	Associated mine name	Type of mine	Estimated amount of coal leased (1,000 tons)[a]	Number Of qualified bids received	Total accepted bonus bid (nominal dollars)	Bonus bid per Ton (nominal dollars)	Bonus bid per Acre (nominal dollars)	Successful bidder
12/15/2004	WYW 151643	2809	Antelope	Surface	194,961	1	$146,311,000	$0.75	$52,084	Antelope Coal Co.
12/29/2004	WYW 150210	2369	North Antelope Rochelle	Surface	324,627	1	$299,143,785	$0.92	$126,254	BTU Western Resources, Inc.
1/19/2005	WYW 154595	2242	Bridger	Underground	32,145	1	$6,953,861	$0.22	$3,101	Bridger Coal Co.
2/16/2005	WYW 151134	2813	School Creek	Surface	327,186	2	$317,697,610	$0.97	$112,959	BTU Western Resources, Inc.
6/5/2007	WYW 160394	1399	Black Butte	Surface	11,200	1	$2,426,620	$0.0002	$1,735	Black Butte Coal Company
2/20/2008	WYW 155132	1428	Eagle butte	Surface	255,000	1	$180,540,000	$0.71	$126,429	Foundation Coal West Inc.
4/22/2008	WYW 174407	2900	Cordero Rojo	Surface	288,100	1	$250,800,000	$0.87	$86,476	Cordero Mining Co.
1/28/2009	WYW 154432	446	Cordero Rojo	Surface	54,657	1	$48,098,424	$0.88	$107,871	Cordero Mining Co.
5/11/2011	WYW 163340	2838	Antelope	Surface	350,263	1	$297,723,228	$0.85	$104,920	Antelope Coal LLC
6/15/2011	WYW 177903	1909	Antelope	Surface	56,356	1	$49,311,500	$0.88	$25,836	Antelope Coal LLC

Table 4. (Continued)

Sale date	Lease serial number	Acres	Associated mine name	Type of mine	Estimated amount of coal leased (1,000 tons)[a]	Number Of qualified bids received	Total accepted bonus bid (nominal dollars)	Bonus bid per Ton (nominal dollars)	Bonus bid per Acre (nominal dollars)	Successful bidder
7/13/2011	WYW 161248	1671	Caballo	Surface	221,735	2	$210,648,060	$0.95	$126,059	BTU Western Resources
8/17/2011	WYW 172657	1024	Belle Ayr	Surface	130,196	2	$143,417,404	$1.10	$140,057	Alpha Coal West
12/14/2011	WYW 174596	1977	Black Thunder	Surface	222,676	1	$300,001,012	$1.35	$151,769	Ark Land Co., South
5/17/2012	WYW 176095	3243	North Antelope Rochelle	Surface	401,831	1	$446,031,864	$1.11	$137,53	6 BTU Western Resources
6/28/2012	WYW 173408	6364	North Antelope Rochelle	Surface	721,155	1	$793,270,311	$1.10	$124,644	BTU Western Resources

Source: GAO analysis of BLM data.

[a] With the exception of WYW 154595 lease tract, we are reporting minable tons of coal for the lease tracts in Wyoming. For WYW 154595 lease tract and lease tracts in all other states, we are reporting recoverable tons of coal. This estimate of mineable tons of coal is generally a larger number than the recoverable estimate, because it includes coal that is generally left in place during actual mining operations, such as coal along property boundaries or coal left in place as pillars for structural reasons in an underground mine. Wyoming BLM does not typically report recoverable tons publicly because officials in this state office consider this sensitive information.

Table 5. Summary Information on Coal Lease Sale Files Reviewed

State	Serial number	Lease tract name	Date of lease sale	Coal offered (millions of tons)[a]	Geologic report prepared?	Engineering report prepared?	Economic report prepared?	Appraisal report prepared?	Fair market value approaches used	Appraisal report signed by three officials?[b]
AL	ALES-047886	Jesse Creek	8/21/1997	0.5	No	No	No	Yes	Comparable sales	Unknown because BLM unable to provide copy of appraisal
AL	ALES-051589	Flat Creek	9/30/2004	10.8	Yes	Yes	No	Yes	Comparable sales Income	No
CO	COC-067232	Dry Fork	1/24/2007	12.1	Yes	Yes	No	Yes	Comparable sales	No
CO	COC-68590	Collom	5/30/2007	92.0	Yes	Yes	Yes	Yes	Comparable sales Income	No
CO	COC-072980	Foidel Creek	1/14/2009	1.4	Yes	Yes	No	Yes	Comparable sales	No
CO	COC-070615	Elk Creek East	5/15/2012	4.0	Yes	Yes	No	Yes	Comparable sales	No
KY	KYES-051002	Gray Mtn.	7/27/2005	2.9	Yes	Yes	No	Yes	Comparable sales Income	No
KY	KYES-050213	Laurel Fork	5/11/2006	0.8	Yes	Yes	No	Yes	Comparable sales	No
MT	MTM-094378	Spring Creek	4/17/2007	108.6	Yes	Yes	Yes	Yes	Comparable sales Income	Yes

Table 5. (Continued)

State	Serial number	Lease tract name	Date of lease sale	Coal offered (millions of tons)[a]	Geologic report prepared?	Engineering report prepared?	Economic report prepared?	Appraisal report prepared?	Fair market value approaches used	Appraisal report signed by three officials?[b]
MT	MTM-097988	Bull Mtn.	2/28/2012	35.5	Yes	Yes	Yes	Yes	Comparable sales Income	No
NM	NMNM-078371	Fence Lake 1	7/31/1991	19.6	Yes	Yes	Yes	Yes	Comparable sales Income	Yes
NM	NMNM-086717	Fence Lake 2	7/31/1991	10.0	Yes	Yes	Yes	Yes	Comparable sales Income	Yes
NM	NMNM-099144	San Juan Deep Ext.	11/1/2000	63.0	Yes	Yes	Yes	Yes	Comparable sales Income	No
ND	NDM-095104	Center Mine	9/12/2006	8.3	Yes	Yes	No[c]	Yes	Comparable sales	Yes
ND	NDM-097633	Center Mine 2	10/15/2009	3.0	Yes	Yes	No[c]	Yes	Comparable sales	No
OK	OKNM-107920	Bull Hill	9/14/2005	9.0	No	No	No	No	No approaches documented	No
OK	OKNM-104763	Liberty West	9/14/2005	2.1	No	No	No	No	No approaches documented	No
OK	OKNM-108097	McCurtain Tract No. 1	9/14/2005	6.5	No	No	No	No	No approaches documented	No
UT	UTU-086038	Miller Canyon	9/3/2009	0.6	Yes	Yes	No	Yes	Comparable sales	No

State	Serial number	Lease tract name	Date of lease sale	Coal offered (millions of tons)[a]	Geologic report prepared?	Engineering report prepared?	Economic report prepared?	Appraisal report prepared?	Fair market value approaches used	Appraisal report signed by three officials?[b]
UT	UTU-85539	Dry Canyon	11/15/2011	42.2	Yes	Yes	No	Yes	Comparable sales	No
WY	WYW-160394	Pit 14	6/5/2007	11.2	Yes	Yes	Yes	Yes	Comparable sales Income Adjusted discounted cash flow comparable sales[d]	Yes
WY	WYW-155132	West Eagle Butte	2/20/2008	255.0	Yes	Yes	Yes	Yes	Comparable sales Income Adjusted discounted cash flow comparable sales	Yes
WY	WYW-174407	South Maysdorf	4/22/2008	288.1	Yes	Yes	Yes	Yes	Comparable sales Income Adjusted discounted cash flow comparable sales	Yes
WY	WYW-154432	North Maysdorf	1/29/2009	54.7	Yes	Yes	Yes	Yes	Comparable sales Income Adjusted discounted cash flow comparable sales	Yes
WY	WYW-163340	West Antelope 2 North	5/11/2011	350.3	Yes	Yes	Yes	Yes	Comparable sales Income Adjusted discounted cash flow comparable sales	Yes

Table 5. (Continued)

State	Serial number	Lease tract name	Date of lease sale	Coal offered (millions of tons)[a]	Geologic report prepared?	Engineering report prepared?	Economic report prepared?	Appraisal report prepared?	Fair market value approaches used	Appraisal report signed by three officials?[b]
WY	WYW-161248	North Belle Ayr	7/13/2011	221.7	Yes	Yes	Yes	Yes	Comparable sales Income Adjusted discounted cash flow comparable sales	Yes
WY	WYW-172657	Caballo West	8/17/2011	130.2	Yes	Yes	Yes	Yes	Comparable sales Income Adjusted discounted cash flow comparable sales	Yes
WY	WYW-174596	South Hilight	12/14/2011	222.7	Yes	Yes	Yes	Yes	Comparable sales Income Adjusted discounted cash flow comparable sales	Yes
WY	WYW-176095	South Porcupine	5/17/2012	401.8	Yes	Yes	Yes	Yes	Comparable sales Income Adjusted discounted cash flow comparable sales	Yes
WY	WYW-177903	West Antelope 2 South	6/15/2012	56.4	Yes	Yes	Yes	Yes	Comparable sales Income Adjusted discounted cash flow comparable sales	Yes

State	Serial number	Lease tract name	Date of lease sale	Coal offered (millions of tons)[a]	Geologic report prepared?	Engineering report prepared?	Economic report prepared?	Appraisal report prepared?	Fair market value approaches used	Appraisal report signed by three officials?[b]
WY	WYW-173408	North Porcupine	6/28/2012	721.2	Yes	Yes	Yes	Yes	Comparable sales Income Adjusted discounted cash flow comparable sales	Yes

Source: GAO analysis of BLM case files for coal lease sales.

[a] In Wyoming, we are reporting mineable tons of coal, while for the other states, we are reporting recoverable tons of coal. This estimate of mineable tons of coal is generally a larger number than the recoverable estimate because it includes coal that is generally left in place during actual mining operations, such as coal along property boundaries or coal left in place as pillars for structural reasons in an underground mine. Wyoming BLM does not report recoverable tons publicly because officials in this state office consider this sensitive information.

[b] According to BLM's handbook on the economic evaluation of coal properties and BLM officials, the three officials that need to sign the appraisal report are: appraiser, reviewer, and deputy state director.

[c] While a separate economic report was not prepared, economic and market information was incorporated into the appraisal report.

[d] This approach involves numerically adjusting comparable sales using the results of the income approach.

For the files we reviewed, table 5 provides information by lease tract on the amount of coal involved in the sale, types of reports prepared as part of the sale, fair market value approaches used, and compliance with appraisal report review requirements.

End Notes

[1] The federal government owns and manages the mineral resources on and below these lands but the state, in cases of state land, or a private party, in the case of private land, owns the surface land. Federal land where coal leasing takes place includes land managed by BLM and other federal agencies, including the U.S. Forest Service. In those instances where a proposed federal lease would be on lands managed by another agency, the federal agency managing the land must consent to offer the lands for lease. In these cases, BLM is still responsible to overseeing the leasing process.

[2] This report uses tons when describing amounts of coal. One ton is 2,000 pounds.

[3] We did not include coal produced from tribal lands in our review of federal coal leasing because they are governed by a different set of regulations and do not involve the same leasing process as the federal coal leasing program. As of December 31, 2012, there were four operations mining tribal coal.

[4] Generally, revenues from federal coal leases are split equally between the federal government and the state in which the coal lease is located.

[5] Another way for companies to obtain the rights to mine coal is through the lease modification process where a company may request a certain amount of contiguous land be added to an existing lease. BLM considers this request and, if granted, the lands are added to the existing lease without competitive bidding. The lease modification process was not part of our review.

[6] More specifically, fair market value is defined as "that amount in cash, or on terms reasonably equivalent to cash for which in all probability, the coal deposit would be sold or leased by a knowledgeable owner willing but not obligated to sell or lease to a knowledgeable purchaser who desires but is not obligated to buy or lease." 43 C.F.R. § 3400.0-5(n) (2013). The Federal Coal Leasing Amendments Act of 1976 require that BLM obtain fair market value for the coal lease tracts and that coal leasing generally be done on a competitive basis.

[7] For the purposes of this report, we are using the term appraisal to mean the valuation of federal coal property offered for lease, in keeping with BLM's guidance. See BLM, *H-3070-1 Economic Evaluation of Coal Properties* (Washington, D.C.: 1994). The value estimated for federal coal lease sales is used to ensure receipt of at least the fair market value as required by the Federal Coal Leasing Amendments Act of 1976.

[8] Before the lease can be issued, the high bidder must also provide a bond to ensure performance of lease conditions, and must undergo an antitrust review by the Department of Justice.

[9] EIA is a statistical agency within the Department of Energy that collects, analyzes, and disseminates independent information on energy issues.

[10] Office of Inspector General, U.S. Department of the Interior, *Coal ManagementProgram, U.S. Department of the Interior* (Washington, D.C.: 2013).

[11] This nonrandom sample cannot be generalized to all coal lease sales held. However, the results of this sample provide illustrative examples of the coal leasing process used and the documentation prepared.

[12] Prior to the enactment of FCLAA, some coal leases were awarded noncompetitively through preference right leases, which were awarded in areas where coal deposits were not known to

exist and were discovered by the applicant. Competitive lease sales were held for coal tracts located in areas with known coal reserves.

[13] GAO, *Analysis of the Powder River Basin Federal Coal Lease Sale: Economic Valuation Improvements and Legislative Changes Needed,* GAO/RCED-83-119 (Washington, D.C.: May 11, 1983).

[14] Two other coal regions were decertified in 1981 and 1982.

[15] BLM has 12 state offices, with most of these located in the western part of the United States. These state offices are located in Alaska, Arizona, California, Colorado, Idaho, Montana, Nevada, New Mexico, Oregon, Utah, Virginia, and Wyoming.

[16] There are instances where the surface is owned by an entity other than the federal government, but the underlying minerals are owned and managed by the federal government. In these cases, known as split estates, the qualified surface owner must consent to any surface mining, but this consent is not required for underground mining. The qualified surface owner does not receive any royalties or other revenues for mining activities but may receive compensation from the lessee.

[17] Enacted in 1970, NEPA has two principal purposes: (1) to ensure that an agency carefully considers detailed information concerning significant environmental impacts and (2) to ensure that this information will be made available to the public. NEPA requires federal agencies to evaluate the likely environmental effects of proposed projects using an environmental assessment or, if the projects would likely significantly affect the environment, a more detailed environmental impact statement evaluating the proposed project and alternatives must be completed.

[18] 43 C.F.R. § 3480.0-5(a)(32).

[19] Some of these state offices oversee leasing activities across multiple states. For example, the New Mexico state office oversees New Mexico, Oklahoma, Kansas, and Texas. There are no federal coal leases in Kansas or Texas according to BLM officials.

[20] There are generally multiple BLM field offices that report to a specific BLM state office. For example, there are 10 BLM field offices in Colorado and 3 of these oversee federal coal leasing.

[21] Heating content is usually expressed as British thermal units (Btu) per pound of coal. A Btu is the amount of energy needed to heat 1 pound of water by one degree Fahrenheit.

[22] Before a lease is issued, the high bidder must also provide a bond to performance of lease conditions, and the lease issuance is subject to a Department of Justice antitrust review.

[23] The Mineral Leasing Act, as amended, directs the Secretary of the Interior to establish annual rentals and royalties for leases but establishes a minimum royalty rate of not less than 12.5 percent of the value of coal recovered by surface mining operations. 30 U.S.C. §207(a) (2013). The regulation establishing the minimum rental rate—43 C.F.R. § 3473.3-1(a)—and the regulation establishing the minimum royalty rate for surface mining—43C.F.R. § 3473.3-2(a)(1)—were issued in 1979. The regulation establishing the royalty rate for underground mining—43 C.F.R. § 3473.3 -2(a)(2)—was initially issued in 1979 with a regulatory minimum (of a 8 percent royalty rate) that could be lowered (to a 5 percent royalty rate) but in 1990 the regulation was amended to establish a 8 percent royalty rate. The regulations also authorize BLM to waive, suspend, or reduce the rental, or reduce the royalty, for the purpose of encouraging the greatest ultimate recovery of federal coal, and in the interest of conservation of federal coal and other resources, whenever it is necessary to promote development or when the lease cannot be successfully operated under its terms, but in no case can the royalty on a producing federal lease be reduced to zero.

[24] EIA. *Annual Coal Report 2011* (Washington, D.C.: 2012).

[25] According to Environmental Protection Agency (EPA) data, coal-fueled electric power plants are among the largest emitters of sulfur dioxide (SO_2) and nitrogen oxides (NO_x), which have been linked to respiratory illnesses and acid rain. The Clean Air Act requires EPA to establish national ambient air quality standards for six pollutants, including sulfur oxides and nitrogen oxides, which states are primarily responsible for attaining. States attain these

standards, in part, by regulating emissions of these pollutants from certain stationary sources, such as electricity generating units In addition, the Clean Air Act Amendments of 1990 established a national cap-and-trade program to reduce SO_2 emissions from fossil-fuel electric generating units and required EPA to establish NO_x emissions limitations for coal-fueled electric power plants. In response to these Clean Air Act requirements, many utilities installed scrubbers and switched to burning low-sulfur coal such as that from the Powder River Basin to reduce SO_2 emissions.

[26] IEA, *Coal Medium-Term Market Report*. Paris, France, 2012. IEA is an autonomous organization established in 1974 that works to ensure reliable, affordable and clean energy for its 28 member countries and beyond. The IEA's four main areas of focus are: energy security, economic development, environmental awareness, and engagement worldwide. Among its key objectives are to improve transparency of international markets through collection and analysis of energy data.

[27] This price per ton of coal is for all coal sold in that state and may include coal from mines that are not on federal lease tracts. This price per ton of coal is the "free on board" price for the coal, meaning it is the price paid for the coal before it is loaded on to a train or barge for transport to its final destination. Thus, this price does not include the cost to transport the coal.

[28] Coal is classified into four major ranks (from highest to lowest): (1) anthracite, (2) bituminous, (3) subbituminous, and (4) lignite.

[29] In this section, the information we present on coal production is based on data from ONRR on the volume of coal sold from federal coal leases. As mentioned earlier, we did not include coal produced from tribal lands in the scope of our report. Thus, the production and revenue information we present does not include tribal lands.

[30] GAO/RCED-83-119.

[31] We included only those tracts where the documentation explicitly said the tracts were modified to enhance competition. We did not include those tracts modified for other reasons, such as to ensure federal coal was not bypassed.

[32] Fifteen tracts were leased after a second sale; two tracts leased after a third sale; and one tract was leased after a fourth sale.

[33] J.A. Luppens, T.J. Rohrbacher, L.M. Osmonson, and M.D. Carter, "Coal Resource Availability, Recoverability, and Economic Evaluations in the United States—A Summary," in *The National Coal Resource Assessment Overview: U.S. Geological Survey Professional Paper 1625–F*, eds. B.S. Pierce, and K.O. Dennen, chapter D (Reston, Va.: 2009).

[34] GAO, *Electricity: Significant Changes Are Expected in Coal-Fueled Generation, but Coal is Likely to Remain a Key Fuel Source*, GAO-13-72 (Washington, D.C.: Oct. 29, 2012).

[35] All dollar figures in this section have been adjusted for inflation to 2013 dollars using the gross domestic price index, unless otherwise noted.

[36] Bonus bid and rent data prior to fiscal year 2003 initially provided by ONRR was limited due to a data system conversion the agency undertook. ONRR ultimately provided updated bonus bid data for this period, but it did so late in our review process, and we were unable to determine its reliability. ONRR provided royalty data for the entire period of our review, starting in 1990, which we assessed to be reliable.

[37] These royalty amounts do not include advance royalties, which lessees, if authorized by BLM, can pay in advance of actual production in lieu of meeting the lease's minimum coal production requirements. In fiscal year 2012, ONRR officials told us that revenue from advance royalties amounted to about $2.1 million or about 0.2 percent of total revenue from coal. If, in years subsequent to paying the advance royalty, the lease meets the minimum coal production requirement, the lessee's royalty will be reduced on a dollar for dollar basis by the amount of the advance royalty. If the lease is relinquished, cancelled, or terminated for any reason, the lessee forfeits any advance royalties paid or due.

[38] An allowance is an allowable deduction from the value of a mineral for royalty purposes. A processing allowance includes reasonable, actual costs incurred by the payer for processing a mineral commodity. A transportation allowance includes reasonable, actual costs incurred by the payer for moving a mineral commodity to a point of sale remote from the lease or unit area, or away from a processing plant; it excludes costs to gather the commodity.

[39] For all types of coal leases, BLM is authorized to reduce the royalty for the purpose of encouraging the greatest ultimate recovery of federal coal, and in the interest of conservation of federal coal and other resources, whenever it is necessary to promote development, or when the lease cannot be successfully operated under its terms, but in no case can the royalty on a producing federal lease be reduced to zero. 43 C.F.R. §§ 3473.3-2(e), 3485.2(c)(1) (2013).

[40] From 1990 to 2011, adjusted average prices for all coal sold in Colorado ranged from $23 to $42 per ton and Utah coal prices ranged from $21 to $35 per ton.

[41] 43 C.F.R. § 3473.3-1(a).

[42] BLM, *H-3070-1 Economic Evaluation of Coal Properties,* I-7.

[43] There is also a third appraisal approach called the cost approach. Under this approach, the value of a property is appraised based on the cost to rebuild or replace the improvements on it. For example, the value of a property with a house on it could be based on the cost to rebuild the house, less any depreciation that has occurred. The cost approach is generally not used to appraise minerals because most of their value is tied to the minerals themselves and not capital improvements.

[44] Adjusting comparable sales using the results of the income approach can be done using the arithmetic and/or proportional approach. In the arithmetic approach, the net present value of the comparable sale is subtracted from the net present value of the tract being appraised. This difference is then added to the actual sale price of the comparable sale. In the proportional approach, the net present value of the tract being appraised is divided by the net present value of the comparable sale, and this adjustment factor is then multiplied by the actual sales price of the comparable sale to adjust it.

[45] The Montana/Dakotas state office manages federally owned minerals in Montana, South Dakota, and North Dakota. There are not any federal coal leases in South Dakota. The New Mexico state office oversees federally owned minerals in New Mexico, Oklahoma, Kansas, and Texas. There are no federal coal leases in Kansas or Texas according to BLM officials.

[46] We spoke with officials from the Appraisal Institute, the Appraisal Foundation, and the American Institute of Mineral Appraisers.

[47] These standards included the *Uniform Standards of Professional Appraisal Practice* prepared by the Appraisal Standards Board of the Appraisal Foundation in the United States, the *Uniform Appraisal Standards for Federal Lands Acquisitions* prepared by the Interagency Land Acquisition Conference in the United States, *Standards and Guidelines for Valuation of Mineral Properties* prepared by the Canadian Institute of Mining, Metallurgy, and Petroleum, *and* the *Code for Technical Assessment and Valuation of Mineral and Petroleum Assets and Securities for Independent Expert Reports* prepared by several groups, including the Australasian Institute of Mining and Metallurgy.

[48] Special Committee of the Canadian Institute of Mining, Metallurgy, and Petroleum on Valuation of Mineral Properties, *Standards and Guidelines for Valuation of Mineral Properties,* (February 2003). These standards are required by the Toronto Stock Exchange, and recommended to be followed in mineral valuations required by regulatory bodies in Canada.

[49] For the other 27 lease sales we reviewed, the initial bonus bid met or exceeded BLM's presale estimate of fair market value for 26 lease sales. For 1 of the 2 lease sales we reviewed in Utah, no bids were received.

[50] Appraisal reports prepared in the Wyoming BLM state office also contained a signature by a fourth official, the branch chief of solid minerals.

[51] Interior, *Departmental Manual*, part 112, chapter 33 (June 1, 2011).

[52] Due to its volume and weight, most coal is transported in the United States by train.

[53] EIA, *Annual Coal Distribution Report* (Washington, D.C.: 2011).

[54] In 2011, coal exports from Montana mines were 13.2 million tons, and coal exports from Wyoming mines were 4.5 million tons. These export amounts may include coal from mines that are not on federal coal leases. EIA, *Annual Coal Distribution Report*.

[55] In 2011, coal exports from states east of the Mississippi River totaled about 84.3 million tons or about 79 percent of total coal exports from the United States. As mentioned earlier, there is little federal coal leasing east of the Mississippi River.

[56] IHS Global Insight is a firm that provides comprehensive economic and financial information on countries, regions, and industries. Wood Mackenzie provides industry and analysis on energy and minerals industries around the world.

[57] Utah BLM officials reported that they were currently preparing for a lease sale and that they would be considering exports as part of this sale. In order to determine the impact of exports on fair market value, the Utah state office contracted with a private firm.

[58] IEA, *Coal Medium-Term Market Report, 2012: Market Trends and Projections to 2017*, (2012).

[59] According to BLM officials, BLM does develop an estimate of the economically recoverable tons of coal for each lease by application for use in valuing each lease.

[60] Coal reserves are different from coal resources. To be classified as reserves, coal must be considered economically producible at the time of classification. Coal reserves are a subset of coal resources.

[61] The Gillette coal field is the largest producing coal field in the Powder River Basin.

[62] James Luppens et. al, USGS, *Assessment of Coal Geology, Resources, and Reserves in the Gillette Coalfield, Powder River Basin, Wyoming* (Reston, VA.: 2008). Spot market prices for Powder River Basin coal were $10.55/ton as of April 2013, according to EIA. Spot prices are prices received on very short-term contracts, generally lasting a few months in length.

[63] USGS, *Assessment of Coal Geology, Resources, and Reserve Base in the Powder River Basin, Wyoming and Montana* (Reston, Va.: 2013).

[64] EIA projects that U.S. coal production will increase at about 0.2 percent per year for the period from 2011 to 2040. If that growth rate continues into the future, estimated recoverable coal reserves would be exhausted in about 194 years if no new reserves are added.

[65] Environmental impact statements are more detailed evaluations of the proposed project and alternatives compared to environmental assessments, and are required if a project would likely significantly affect the environment.

[66] For environmental assessments a decision record is used; for environmental impact statements a record of decision is issued.

[67] OMB Circular No. A-130.

[68] GAO, *Agricultural Chemicals: USDA Could Enhance Pesticide and Fertilizer Usage Data, Improve Outreach and Better Leverage Resources*, GAO-11-37(Washington, D.C.: Nov. 4, 2010), and GAO, *Medicare: Communications to Beneficiaries on the Prescription Drug Benefit Could Be Improved*, GAO-06-654 (Washington, D.C.: May 3, 2006).

[69] BLM, *H-3070-1 Economic Evaluation of Coal Properties*, V-4.

[70] BLM officials noted that federal officers and employees are prohibited from publishing or disclosing proprietary or business confidential information to any extent not authorized by law and that unauthorized publication or disclosure could result in criminal penalties.

[71] BLM, *H-3070-1 Economic Evaluation of Coal Properties,* V-5.

[72] Pub. L. No. 89-487 (1966), *codified as amended at* 5 U.S.C. § 552. FOIA requires federal agencies to provide the public with access to government records and information on the basis of the principles of openness and accountability in government.

[73] The other requester did not appeal BLM's decision.

[74] The Eastern States office oversees activities in the eastern half of the United States. There is currently coal leasing activity in Alabama and Kentucky. The New Mexico state office also oversees leasing activity in Oklahoma.

[75] We reviewed three pre-2007 files for both Oklahoma and New Mexico because these sales involved multiple lease tracts that were held on the same date.

[76] One of the two lease sales we reviewed from Utah did not receive any bids. All of the other lease sales received at least one bid.

[77] We reviewed three pre-2007 files for both Oklahoma and New Mexico because these sales involved multiple lease tracts that were held on the same date.

INDEX

G

H

I

J

L

M

N

O

P

Q

R